Water Sensitive Cities

Water Sensitive Cities

Carol Howe and Cynthia Mitchell
Editors

Publishing
London • New York

Published by **IWA Publishing**
Alliance House
12 Caxton Street
London SW1H 0QS, UK
Telephone: +44 (0)20 7654 5500
Fax: +44 (0)20 7654 5555
Email: publications@iwap.co.uk
Web: www.iwapublishing.com

First published 2012

British Library Cataloguing in Publication Data
A CIP catalogue record for this book is available from the British Library

Library of Congress Cataloging-in-Publication Data
A catalog record for this book is available from the Library of Congress

ISBN: 9781843393641
ISBN: 1843393646

Contents

Foreword

Rob Skinner

Professorial Fellow, Centre for Sustainable Cities, Monash University, Australia.

Over the past two years a series of water industry workshops internationally[1] has explored the role that water plays in delivering liveable, sustainable, and productive cities.

In the southern states of Australia this has been a particularly critical, strategic exercise given the dramatic changes that have occurred in our climate. In Melbourne, for example, the average streamflows into our major reservoirs have reduced by 37% over the last fifteen years and the response to this significant drop in water yield has been to construct some major non-rainfall dependent sources of water, the most significant being a very large desalination plant. However, the lesson we have learned from this experience is that sustainable cities of the future will need to have in-built resilience to shocks from further changes in climate, population increases, financial crises or major natural disasters. And the secret to building resilience into a city starts with the way the city is planned and designed from the outset, to ensure that the city itself is 'water sensitive'. Diversity of solutions is a key. The urgency to make these fundamental change is why we in Australia have been so committed to the Cities of the Future movement and the workshops and conferences that have taken place in recent years.

[1]Workshops included Singapore Water Weeks 09 & 10, SWITCH CityWater Summit (Delft), SWIF Workshop (Beijing) OzWater10 (Brisbane), Enviro2010 (Melbourne), International Water Association (IWA) World Water Congress 2010 (Montreal), WEF/IWA Cities of the Future (Boston). American Planning Association (San Francisco) and UC Berkeley events (New Orleans)

The strong conclusion from these workshops is that water has a critical role to play in achieving important outcomes in these areas:

Liveable Cities – by delivering safe, fit-for-purpose water supplies; attractive urban landscapes that support healthy communities and improved flood protection;
Sustainable Cities – by ensuring smaller environmental footprints; healthier waterways and parklands; landscapes that are resilient to natural disasters and climate variability;
Productive Cities – by providing water security for the future; affordable water services; a clear, transparent and contestable investment climate; and economic prosperity.[2]

Cities that achieve these water-related outcomes will be what the International Water Association (IWA) refers to as 'Cities of the Future'. In so doing these cities will have adopted water sensitive planning and design principles. The contributors to this book are all working within these principles or exploring how to achieve them.

But what specifically are these Cities of the Future principles?

The 2010 workshops referred to earlier attracted a wide range of professional planners and policy makers (over two hundred in all) from both the water sector and non-water areas of city planning. The primary aim of this series of workshops was to develop a shared 'Vision' and subsequently a suite of 'Principles' that would govern the development of policies and strategies for Cities of the Future.

As a result of this work the IWA has endorsed twelve Principles[3] under four Themes (with a suggested work program to progress their implementation).

The four themes are:

1. *Cities of the Future will be liveable and sustainable.* A lot is said about the ideas of 'liveability' and 'sustainability' – who could argue against such worthwhile ideals – but the Principles help to clearly define the concepts in terms of the city as a whole and how water links to them.
2. *The many values of water* – whilst this is a more mainstream technical theme of the IWA, it takes on a new life if we view all components of the water cycle (even the stormwater and 'waste water' components) as a potentially valuable resource for the city.

[2]Living Victoria Ministerial Council 2011, *Living Melbourne, Living Victoria Roadmap*, DSE, Melbourne

[3]Binney P, Donald A, Elmer V, Ewert J, Phillis O, Skinner R, and Young R 2010, Cities of Future Program, Spatial Planning and Institutional Reform Conclusions from the World Water Congress, Montreal, September, 2010

3. *Community choice and knowledge sharing* – timely and reliable information and open knowledge sharing are the essential prerequisites for integrated and sound decision making.
4. *Adaptive and collaborative water sector* – water has a central role to play in achieving water sensitive Cities of the Future – but this will only be achieved if the water sector has its thinking and desired outcomes integrated with other city planning processes from the outset. Transitioning to this state is a dominant theme of this book.

The Principles under each theme are presented below:

Theme 1: Liveable and Sustainable Cities

Principle 1: **Cities will continue to grow in population but will be increasingly liveable. A feature of desirable Cities of the Future will be more interconnected communities.**

Cities are complex, dynamic systems that are likely to become more complex over time. Cities will continue to offer lifestyles – jobs, cultural attractions, recreation and sporting attractions – that will attract people in abundance. Principle 1 recognises that people value a liveable city that provides the amenities and space to maintain local connections and healthy communities.

Principle 2: **Cities of the Future will provide access to safe drinking water and sanitation for all.**

The United Nations Development Program estimates that currently almost 1 billion people lack clean drinking water, while 2.4 billion people have no access to hygienic sanitation facilities and 1.2 billion lack any sanitation facilities at all. Although people in developing nations account for most of these statistics, there are also sections of the population in developed nations that lack these basic services. While the technologies exist for providing low-cost water and wastewater services, effective water governance is the missing link to achieving more equitable water resource management and service delivery.

Principle 3: **Sustainable Cities of the Future will combine a compact footprint with sustainability and liveability.**

Cities of the Future will need match higher-density living with 'green urban design', and link spaces to provide the ability to easily connect with other parts of the city. Lower-density living will also be available within the city to provide a range of living options.

More water-sensitive cities will be greener and, therefore, cooler. With lower 'urban heat island' effects (the tendency of urban areas to be hotter than their more vegetated surroundings), these cities will be healthier places in which to live.

Principle 4: **Cities of the Future will be resource-neutral or generative, combining infrastructure and building design that will harmonise with the broader environment.**

The urban form will generate water, energy and nutrient by-products that can meet the city's resource demands in a way that is carbon neutral. Some cities may generate resources in excess of their needs and be able to supply demands in surrounding regions.

Cities will also be designed to operate in harmony with the broader environment. For example, cities will release water to the environment consistent with natural environmental flow patterns.

Principle 5: **Sustainable cities will be part of prosperous, diverse and sustainable regions.**

Cities will not function as isolated entities. Instead, they will be interdependent with their regional partners, respecting 'local identity' and valuing the flow of resources, people and information between the two.

Cities themselves will enjoy prosperous economies built on sustainable communities, and its citizens will act to bring out the best in themselves and their surrounding regions.

Theme 2: The Many Values of Water

Principle 6: **Sustainable cities will be served by a well-managed water cycle that, in addition to public health and water security, provides for healthy waterways, open spaces and a green city.**

Water will be managed across the water cycle and watershed to deliver economic and social value for the community, and to protect and enhance environmental values and biodiversity.

Principle 7: **Sustainable cities will recognise that all water is good water – based on the concept of 'fit-for-purpose' use.**

It will be recognised that water has many different values and 'fit-for-purpose' uses. All water comprising the urban water cycle (including stormwater and wastewater) will be highly valued and managed to deliver optimal environmental and social outcomes.

Theme 3: Community Choice and Knowledge Sharing

***Principle 8:* Cities of the Future will be served by informed, engaged citizenry and multi-scale governance that enables local community choice.**

Communities place greater value on their resources where they have greater control over them. On this basis, water will be valued and utilised best when its users are informed and able to exercise appropriate levels of local choice.

Communities will choose the future of their cities and the way that they live in these spaces. They will choose the pathways that they take to get to reach these goals.

***Principle 9:* Customer sovereignty with full environmental and social cost.**

Citizens – as customers and developers – will be able to pursue their individual choices, while ensuring sustainable outcomes, by bearing the full environmental and social cost of those choices.

Being fully informed and bearing the full costs of their decisions will prompt businesses and individuals to demand efficiency and affordability in the actions that shape water consumption (for example, water-sensitive urban design in the case of builders and developers, recycled water systems, water-efficient appliances).

Citizens will have a well developed sustainability ethic that informs all of their decisions. This will be backed up with appropriate tariffs, transfers and taxes that encourage good behaviour. Proper alignment of economic incentives and environmental regulation is essential for creating Cities of the Future.

***Principle 10:* Accurate and useful information, including smart metering.**

Informed citizen choice depends upon full knowledge of the resources available, the potential benefits of different options and the evaluation of on-going performance.

Cities will draw more fully on intelligent information and management systems across a full range of networks, including smart water system design to provide information to system managers and users. These systems will synthesise data from across the water cycle and share it across utilities and customers to inform decision making.

Theme 4: Adaptive and Collaborative Water Sector

***Principle 11:* Sustainable cities will be served by adaptive and integrated approaches to urban development.**

Sustainable Cities of the Future will be realised when the sectors that supply services to cities work more closely with governments, planners, businesses and the community from the first stages of urban planning.

Given the linkages between water, city shape and design and energy consumption, a transformation in these and other sectors to more integrated planning will underpin the development of resilient cities in the future. This integration will occur at all scales of planning.

> ***Principle 12:*** **Sustainable cities will be served by a multi-faceted water management system.**

The water sector will become more diverse and dynamic, drawing on integrated solutions within the water sector, across sectors and including government and the community.

Some water providers may diversify to become multiple utility providers. Others may become total water cycle providers, and others still may enter the sector to provide a mix of public and private service providers.

CONCLUSIONS

It has been a significant step for water sector planners and engineers to collaborate with their counterparts in other sectors to develop the IWA Principles for Cities of the future.

To agree on Principles such as 'engaged citizenry', that are not linked directly to engineering processes, is an evolution from what IWA members have traditionally regarded as their core business.

And yet these are the areas that the water sector engineers and planners must influence.

Cities of the Future as described by the IWA's Principles are proposed as a response to our current water security challenges that is also an essential component of building liveability and sustainability into our cities.

This approach involves moving beyond the traditional practice of only supplying and removing water from a city from centralised sources. It considers where the water comes from (for example, alternate sources), how it is used within the city (for example, water supply, urban heat island mitigation, sporting field irrigation, market gardens) and where it goes (for example, reducing stormwater waste).

Within this approach decisions are not only technical–they also involve people and land use consideration to embed decentralised options into the city and demonstrate the value they can provide to customers.

Achieving a City of the Future therefore relies on influencing the urban form, and involving customers and community in water management decisions in addition to the traditional array of technical experts.

It is within this context that this book is both a timely and important contribution to the emerging theory and practice surrounding Cities of the Future.

The contributors have worked extensively either within or in close collaboration the water sector. They are presenting insights and directions that are directly relevant to the new Cities of the Future Principles – putting their work into themes such as 'winning the minds of the community', 'building water into urban design' and 'transitioning to a water sensitive city'.

If the process of developing these principles taught us anything, it is that 'collaboration' and 'engaged citizenry' are the new currencies of city planning, and providing leadership in the transition to fully integrated planning is now a major imperative.

Acknowledgements

Many people have contributed to making this book a reality. The encouragement of Harry Seah (Singapore PUB) and Darryl Day (Northern Territories Power and Water, Australia) to develop exciting sessions for the Planning for Sustainable Water Solutions Conference Theme during the 2009 Singapore Water Week gave the impetus for this book to be created. Steve Moddemeyer (Collins Woerman) for his help in selection of speakers and papers for the event. Paul Reiter (IWA) for his vision and leadership of the IWA Cities of the Future iniative. Kala Vairavamorthy (University of South Florida) and Paul Brown (CDM) for providing leadership in taking the programme forward. The many authors who contributed to this book. It's a significant effort to translate a conference presentation into a chapter, and moving forward at the rate we need to requires us to learn from each other – that only happens if people have the generosity and commitment to make the time to share their learnings in forms like this book. Maggie Smith and Michelle Jones from IWA publishing for their advice, editorial support and assistance in publishing the book. Also, for their on-going patience in delivery of the final versions of the chapters. Finally, we would like to acknowledge the many researchers and practitioners across the globe who are actively investigating and implementing new system solutions that embrace the principles espoused by the IWA Cities of the Future programme.

Chapter 1

Introduction

Cynthia Mitchell[1] and Carol Howe[2]

[1]*University of Technology Sydney, Institute for Sustainable Futures, Sydney, Australia*

[2]*UNESCO-IHE Institute for Water Education, Delft, The Netherlands*

There are two kinds of change – incremental improvements and radical, step changes. Right now, across the globe, the urban water sector is undergoing a seismic shift. A new generation is emerging with a new set of priorities and principles, and Singapore International Water Week in 2009 was a watershed event in that journey. This book had its genesis there. Our goal with this book is to help propel the shift by showcasing the leading edge of thinking and practice from around the world.

Our work at the Institute for Sustainable Futures has been guided for some years by the conceptual framework of generational shift (see White (2005) for the original idea and Abeysuriya et al. (2007) for further description). As Figure 1.1 shows, the earliest phase of water and wastewater service provision can be characterized as 'unmanaged'. At this stage, the financial cost to householders is low, the environmental burden is local, and the public health outcomes are likely to be poor. Typically, centralized service provision comes next, providing significant improvements in public health outcomes, increasing the financial cost, and shifting the environmental burden to a regional location through, for example, dams and discharges. The early days of water cycle thinking are well-intentioned, driven by goals such as maximizing recycling, or maximizing nutrient removal. However, the outcome of investments in this generation increase both the cost and ecological impact of service provision and shift the environmental burden still further afield, for example through increased energy intensity and greenhouse gas emissions. In contrast, the fourth generation takes a more sophisticated approach that draws on the concepts of natural capitalism made famous by Paul

Hawken, Amory Lovins and Hunter Lovins, (Hawken et al., 1999) and seeks to 'tunnel through the cost barrier', finding drastic improvements in material intensity to arrive at restorative outcomes – outcomes that do more than minimize harm (Mitchell 2008).

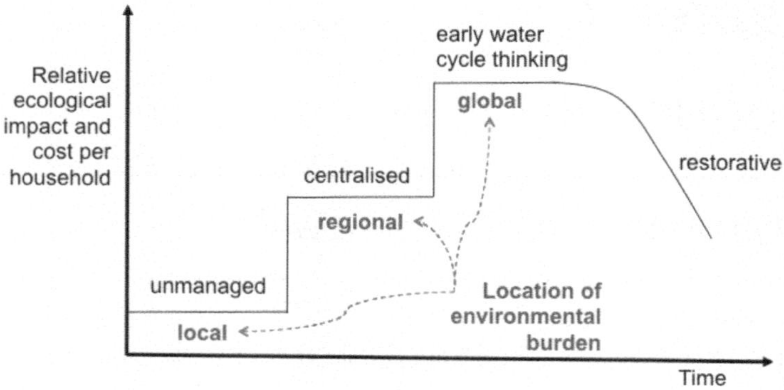

Figure 1.1 Generational change in the water sector

Transitioning to this new generation requires a great deal of learning, both in terms of what we need to do (principles and frameworks) and how we go about doing it (practices). This book seeks to bring together varied contributions in both of these realms, firstly to demonstrate that both are necessary – top down principles and bottom up practices, and secondly to show that there are many possible paths and what matters is to integrate the framework and the practice with the local conditions.

The first set of chapters frame the transition from different perspectives (Nelson; Brown; Mitchell, Fam and Cordell; Kenway and Lant; and Binney). The second set of chapters set out the need for and successful examples of engaging broadly, deeply, and differently with diverse groups, including water users, as a means of ensuring successful uptake of new approaches (Howe; Simpson; MacPherson; Guan and Toh; and Harley). The third set of chapters explore the principles of assessment that are needed to guide design and re-design (Novotny and Novotny; van der Steen). The final set of chapters put principles into practice, and show how people and cities are already experimenting with new and existing settlements across the globe, from harsh desert environments to delta cities (Jacobs; Elmer and Fraker; Hartman, Kulaib, and Witherspoon; Wagner and Zalewski; de Graaf and van de Ven). Below, each chapter is briefly introduced.

Valerie Nelson (USA) sets the scene for the book, introducing a new paradigm based on the 'water commons': a smart, clean, green approach that recognises our current practices are not financially, ecologically, or societally sustainable. Nelson notes that whilst achieving an optimal mix is difficult, it is also urgent, so she calls us to 'act artfully now', and provides a plethora of suggestions for how we might do that, including lessons from other sectors.

Rebekah Brown (Australia) brings a theoretical perspective and structure to the space, drawing on transition management to outline how the industry's old assumption of stationarity is incapable of responding to the uncertainty we are faced with, and proposing a 'water sensitive cities' framework as an adaptive response that sets up the requisite breadth in the 'hydrosocial contract' between water service providers and water users.

Cynthia Mitchell, Dena Fam and Dana Cordell (Australia) set out the characteristics of restorative futures, and argue for transition management concepts as a useful guide for making the shift. To ground the ideas, they explore the case of phosphorus scarcity and security, and demonstrate the need for new governance arrangements.

Steven Kenway and Paul Lant (Australia) focus on urban metabolism as an overarching framework. They track the rise and rise of energy intensity of water services, and set out the need for far greater attention to the water energy nexus. Their chapter is a wonderful resource of rich information, and a plea for better analytical frames that allow real assessment and comparison of alternatives from an energy and carbon perspective.

Peter Binney (Australia/USA) draws on decades of experience of working closely with the water sector to distil lessons learned for sustainability, calling on the industry to position environmental protection within a broader frame of financial stewardship and community development.

The next set of chapters is focused on the need to go well beyond a technical focus, and to invest in learning from and bringing the people with you. Carol Howe (Netherlands) explores how transitions need real diversity from a people perspective, across values, roles, and disciplinary qualifications.

Jenifer Simpson (Australia) brings a much-needed outsider's view, showing how our arcane language within the industry makes it difficult to communicate effectively with our communities, and explaining that the vacuum in community knowledge leaves water management open to political exploitation, especially when it comes to recycling. With gentle humour, she goes on to point out anomalies aplenty in how the water sector goes about its business, including the very mixed message about how unplanned recycling happens all the time and is fine, but planned recycling is apparently more risky because it requires a separate set of guidelines and regulatory arrangements. Linda MacPherson (USA/Singapore) makes a similar plea from an industry perspective, exploring why reuse has often failed, and noting that existing water reuse is the industry's best kept secret. She goes on to provides guidance from case studies from around the world about how to improve outreach, terminology, imagery and experiences for users.

Yap Kheng Guan and Sally Toh (Singapore) chart the journey of Singapore's extensive and successful efforts to bring their 4.5 million residents on board with a diverse set of water sources, including significant water recycling and what they term 'NewWater' and 'indirect potable use', bringing to life the idea of focusing on the quality of water provided, rather than its history. Brendan Harley (USA/Singapore) completes this set and acts as a segue to the examples with the extraordinary story of Singapore's Marina Barrage, a true multi-purpose project that brought engineers into close encounters with the public, and has resoundingly met three very different goals: a tidal barrier and flood control; a very urban reservoir; a lifestyle attraction. Less than two years after opening, the barrage welcomed its millionth visitor – outstanding in such a small country.

The third set of chapters focus on assessment, Vladimir and Eric Novotny review a broad group of current and future eco-cities, from Masdar to Hammarby Sjostad via China and the USA, and conclude that a critical gap is macroscale measures, models, and indices to assess these initiatives. On the other hand, Peter van der Steen draws on the extensive experiences from the international SWITCH network to show how the combination of local, system-wide strategic planning and particular assessment frames gave clear and useful guidance for existing cities, for example, applying quantitative microbial risk assessment to the whole water system in Accra, Ghana, showed that the best investment to improve public health outcomes was actually in sanitation, rather than further improvement of water supply which is the usual response.

The final set of chapters is practice-based, either in existing settlements or in designing and constructing new settlements. All show a strong linkage and commitment to sets of principles akin to those set out by Rob Skinner in his Foreword. This should not be surprising, since the Cities of the Future movement has been gathering pace internationally at the same time as this book has been in production. Iwona Wagner and Maciej Zalewski (Poland) show how ecohydrology principles helped deliver better quality of life outcomes in creating a blue-green network as part of transforming Lodz from an industrial centre to a liveable city. Vicki Elmer and Harrison Fraker look at six functioning eco-city neighbourhood scale developments across Europe to show how integrating water, energy and solid waste functions provides feasible micro-utility opportunities and significant sustainability synergies, particularly when the landscape is invoked as the 'fifth infrastructure'. John Jacobs (Netherlands) recounts how Rotterdam is embracing its water-based constraints and turning them into opportunities through its approach to climate proofing, seeking to position itself as a market and knowledge leader whilst improving local neighbourhood quality. Similarly, Abu Dhabi sees Masdar, its brand new sustainable city for 40 000 residents and 50 000 daily visitors and workers, as a major positioning opportunity. Meghan Hartman, Kulaib and Witherspoon step through the modern-day challenges and opportunities to learn from ancient insights in designing this zero-waste philosophy city in the harsh climate of the Middle East. Finally, de Graaf and van de Ven use initiatives from delta cities in the Netherlands and Japan to show how a balance of stakeholder receptivity and spatial planning and development can facilitate sustainable, climate robust, adaptable, healthy, and pleasant water-based cities.

To sum up the book and the introduction, we have used Wordle™. Wordle™ does a word count and represents the outcome as a graphic such that the more frequently the word is found in the text, the bigger it is in the graphic. Figure 1.2 is a Wordle™ of this book. What it clearly shows is that our focus is on cities and urban water, and that in this context, energy is of primary importance as we look to the future, as is thinking in terms of systems. And in this particular Wordle™ representation, phosphorus is the last word, which seems appropriate given Mitchell et al's (this volume) prediction that phosphorus will drive massive change in our sector in the medium term.

Figure 1.2 A Wordle™ representation of the themes in this book

REFERENCES

Kumudini Abeysuriya, Cynthia Mitchell, and Stuart White (2007) Can corporate social responsibility resolve the sanitation question in developing Asian countries? Ecological Economics 62:174–183.

Paul Hawken, Amory Lovins, and L. Hunter Lovins (1999) Natural Capitalism: Creating the Next Industrial Revolution, Earthscan.

C. A. Mitchell (2008) 'Restorative water: beyond sustainable.' Waste Management and Environment Magazine July 2008 pp. 20–21.

S. White (2005) 'The coast, the dam and in-between: issues for Manly's future', Manly Futures Forum, Manly Council, September 2005. View/Download from: http://www.isf.uts.edu.au/publications/white2005manlyfuturesforum.pdf.

Part I
Framing the Transition

Chapter 2

Achieving the water commons – the role of decentralised systems

Valerie Nelson

Water Alliance, Gloucester, Massachusetts, USA

ABSTRACT

Traditional water management has relied on an industrial-scale engineering and economic model developed in the 1800's. With a goal of public health protection, big pipe systems were built to transport clean water in and wastewater out of urban neighborhoods. Low water and sewer rates have been based on extraction of water as a free natural resource, moving water from one basin to another, moving mass volumes of fresh water into brackish and salt water environments, and dumping wastes into the environment with minimal penalty. The damage from these approaches is manifested in eutrophied lakes and estuaries, falling levels of groundwater and streamflows, loss of habitat, absorption of toxins by humans and other organisms, and reductions in public health and community well-being, particularly in low-income, marginalized communities. An emerging water paradigm relies instead on design principles found in nature: in particular, integrated systems, efficiency and reuse, and adaptation to local conditions. These 'smart, clean, and green' approaches create a wealth of services and benefits at the local level in cities and towns and can help restore the ecological and societal well-being of the 'Water Commons' as well.

Keywords: Decentralization; Integration; Water Commons; Multiple Benefits; Resource Recovery; Cities of the Future

2.1 INTRODUCTION

Conventional systems of water management have been considered one of the 20th Century's greatest public health accomplishments. Large networks of water and sewer lines and treatment plants brought clean water into the cities and transported away disease-carrying sewage. With passage of the 1972 Clean Water Act, the United States also established a goal of fishable, swimmable surface waters and funded and enforced increasingly stringent regulatory standards for partially removing pollutants from treatment plant effluent. The 1974 Safe Drinking Water Act focused on setting up engineering barriers to drinking water sources but in doing so ramped up the cost of providing water treated to a potable standard for many non-potable uses. Municipal utility management was the norm. Rural towns with private wells and septic systems were expected over time to build public water and sewer systems as well.

In recent years, however, a concern has been growing that this 'paradigm' of big-pipe water management is not sustainable, both from a natural resource and a financial perspective. The appropriation of huge volumes of water from natural systems and the release of polluted effluent into rivers and the oceans have been increasingly disruptive of ecosystems. Signs of natural resource stress are seen in falling groundwater levels and decreasing stream flows, eutrophication of lakes and estuaries, disappearance of wetlands, dead zones in coastal areas, and other changes in hydrological function. A less well known but potentially even more significant sign of stress is the component of the water cycle that is mediated through vegetation: evapo-transpiration. Massive reductions in vegetation, and therefore massive reductions in water entering the atmosphere through evapotranspiration, are now thought to be a substantial contributor to global warming (Kravcik et al., 2008).

Signs of financial stress are evident in the functionality of infrastructure across the US and elsewhere. For example, drinking water systems lose considerable amounts of water from leaking pipes, treatment technologies cannot keep up with emerging biological and chemical contaminants, and treating all water to new and more stringent standards is both increasingly difficult and, one could argue wastes valuable energy, chemicals and money, except for the small amount of water needed for potable uses. Some cities and towns in the US have been unwilling to charge ratepayers the full cost of repairing and replacing the infrastructure, and so collapsing pipes and breakdowns in treatment plants have become more frequent (Cooper, 2009).

What is clear is that we need new, radically different, models to guide both the rejuvenation of our existing infrastructure, and the provision of new services, towards systems that are sustainable in ecological and economic terms. Section 2 describes just such a new approach that mimics and works with nature, and sets out the need for and role of decentralized systems within broader networks. Section 3 introduces the foundations of a new framework for moving forward – the concepts of a 'Water Commons', 'water at the heart of all life', and 'smart, clean, and green' design. Section 4 steps through the required design practices and values, and includes examples and thought experiments that show what is necessary and possible. Finally, Section 5 identifies the constraints of our current water governance arrangements, and explores ideas, including learnings from other sectors, for the shifts in policies that are necessary to achieve lasting, long-term change towards a Water Commons for all, in the USA and elsewhere.

2.2 INFRASTRUCTURE THAT MIMICS AND WORKS WITH NATURE

The 2007 Baltimore Charter for Sustainable Water Systems asserts an alternative approach to water management that 'mimics and works with nature' (Nelson, 2007). Natural systems create an abundance of value and diversity, where species cooperate and one species' waste is another species' resource. The genius of science and design in the 21st Century is in unlocking the complex patterns and processes of nature and of applying these lessons to 'smart, clean, and green' infrastructure and buildings.

In practical engineering terms, this new approach includes: provision of potable-quality water for drinking water and direct human contact purposes only; prevention of pollution before it gets into the waste stream (including the wholesale re-engineering of some products through green chemistry to mitigate or eliminate ecological damage); reduction of energy needs by avoiding the pumping and long-distance transport of water and wastewater; wastewater recycling and non-potable, 'fit for purpose' reuse instead of disposal; rainfall harvesting; energy and nutrient recovery from wastewater; habitat and natural system restoration; re-vegetation to restore evapo-transpiration capacity; and restoration of green infrastructure in urban areas to help beautify cities and revitalize neighborhoods.

2.2.1 Networks of decentralized and centralized infrastructure

A birds-eye view of the new infrastructure would reveal 'networks' of decentralized and re-purposed and at times hybridized, centralized systems. These engineered and green networks mimic the natural systems of nodes and links in nature, where water both recycles and supports life at a local scale, but also is a linkage and transport mechanism across a landscape and with the atmosphere.

Some of the innovative treatment and resource recovery technologies would be 'embedded' in subdivisions, apartment complexes, or individual homes and offices. Other functions would be taken over by vegetative 'green infrastructure', such as green roofs and walls, trees and swales along roads, and restored streams and wetlands. Water and sewer lines might be slip-lined and re-purposed for potable or reclaimed water, storage, and heat recovery. Monitoring and control technologies would be key elements in managing the system and in protecting public health. Incentives and regulatory mandates will be required to ensure that these sustainable practices are adopted and that the remaining watershed and global 'externalities' are also addressed by developers, homeowners, industries, and municipalities.

It is difficult to estimate the costs of systems like this in cities and towns, and more difficult still to compare such costs with those of existing approaches, not least because the full costs of existing approaches are at best unclear and partially, unintentionally, hidden through subsidies, sunk costs, and the like. It seems likely that the full cost of these new approaches, once mainstreamed, would be no more than the full cost of current approaches. What the new approaches offer instead is an extraordinary array of new potential benefits, in terms of air quality, energy savings and production, recreation, beauty and aesthetics, increased property values, and jobs.

The public in the US is increasingly familiar with the concept of decentralized/centralized networks through the shifts in the energy sector, where a transformation to a distributed and efficient network that relies in significant part on clean natural system services is underway. The existing power grids use large electrical networks and power plants, as well as oil and gas pipelines, that deliver energy to homes and businesses at subsidized rates and produce large externalities in air pollution and greenhouse gas emissions. In contrast, the new grid will incorporate clean natural sources of energy, including wind and solar, at distributed locations and will encourage more energy-efficient building designs. Alternative energy sources are the equivalent of 'fit for purpose' water sources. Metering and incentives for peak generation and off-peak use of energy are just as important for water provision. 'Smart grids' also rely on sophisticated information and control systems. Both energy and water networks require supportive financial incentives and regulatory mandates.

2.2.2 Interdisciplinary integration across infrastructure sectors

Leading-edge infrastructure experts are now going further, as the examples in this book show, and suggesting not only integrating these 'networks' of engineered and green energy and water systems, but also extending the integration and co-engineering to include transportation, solid waste, buildings, and other urban infrastructure (NRC, 2009). The lessons of nature suggest that such integration will lead to further significant synergies of design, cost-savings, and an abundance of positive benefits for society. For example, an 'eco-block' incorporating architectural innovations, wind and solar power, green roof and wall cooling, rainwater harvesting, water reuse and energy recovery, and nutrient recycling into community gardens, can be nearly 'off-the-grid' in both energy and water, and can be located at transportation 'hubs' (Fraker and Wurster, 2008). These new infrastructure designs have the potential to both improve the quality of life in urban communities and begin to protect and restore the ecological commons.

Finally, the solutions to water management in the 21st Century will require a high level of interdisciplinary collaboration and broad public engagement. Here also, nature serves as a model for the benefits of collaboration and cooperation in society, as opposed to the specialization and hyper-individualism of the 20th Century. Networks of conversations and pilot projects will serve as the foundation for creative invention and enhancement of the 'Common Wealth'.

2.3 RESTORING THE WATER COMMONS

One of the most significant barriers to the full and appropriate use of decentralized water resource systems is the failure to consider externalities of central system approaches. Siloed utilities and individual communities are allowed, subsidized, or forced by court decrees to proceed with highly-disruptive individual projects that cumulatively cost more and create greater environmental, social, and economic damage than an approach which both decentralizes and integrates various water resource sectors across a region.

For example, construction of water lines to provide safe drinking water typically leads to increased water use and eventual failure of septic systems. Construction of sewer systems to deal with these problems promotes further housing development and an increase of polluted stormwater runoff, along with groundwater depletion if wastewater is released into surface waters and there is not adequate infiltration. An integrated water, wastewater, stormwater and development plan from the start could both save money and better protect surface water quality and groundwater supplies.

Garrett Hardin's 'Tragedy of the Commons' (Hardin, 1968) is a useful framework for understanding the problem of externalities. Hardin's model was the over-grazing of community land (the British village green or 'Commons'). The individual herdsman had no incentive to cut back the grazing of his own cows, since others would just take his place. But, cumulatively and over the long-term, the land would degrade and collapse in its capacity to support grazing for local herdsmen. One solution to this 'Tragedy' was to develop community management systems that restricted effort, another to privatize the land so that each owner would internalize the costs of over-grazing and reduce effort so as to preserve eco-functions on his own land.

In water and water-related sectors, a concern is that a global 'Tragedy of the Commons' may be on the horizon as well. As Jared Diamond has pointed out, past civilizations have turned lush landscapes into deserts through profligate use and mismanagement of water (Diamond, 2005). The new literature on 'resiliency' suggests there are tipping points in nature, where ecosystem collapse is difficult if not impossible to reverse (Walker and Salt, 2006). For example, a lake can absorb and store some nutrient pollution, but at some point the lake 'dies' and can't be cleaned and revitalized at a reasonable cost. There are similar concerns for tipping points in the capabilities of microbial populations in seas and oceans to absorb carbon and acids (Hough-Gulber, 2007).

2.3.1 Water at the heart of all life

Water is connected to so many ecosystem and societal functions and services, from the elemental to the spiritual, that the phrase is often used 'water is at the heart of all life.' Water's many connections to the 'Commons' may be stressed in the future to the point of collapse. These losses include depleted water supplies for potable, commercial and industrial uses, desertification, drying and degradation of soils, depletion of phosphate stores needed for growth of plants, toxic overload in the environment, energy use and disrupted evaporation cycles contributing to climate change, and others.

There are numerous other impacts on the Commons that may fall short of 'collapse', but that nevertheless represent a degraded quality of life on Earth. The 'Commons' is a useful term to describe the space in which natural organisms, including humans, interact. There is an ecosystem web of life with complex interdependencies of species and natural resources of soils, energy, and water. There are also complex interdependencies of people in human societies and economies, which in turn rely on nature's services for their survival. The Commons can be used to describe that space of interactions and interdependencies that affect the well-being of all the individual participants, both human and non-human.

This is a particularly interesting and critical moment in history, when both the ecological and societal Commons are threatened. Climate and other natural systems are under increasing stress globally, but so are communities and the economy. New discussions are being held about the need to restructure relationships in the Commons so that ecological, social, and economic systems are together restored to health (World Business Council for Sustainable Development, 2010).

In 1969, Ian McHarg described poignantly in Design With Nature the graying and pollution of cities and towns, with tremendous loss of physical and emotional well-being of residents (McHarg, 1969). Water mismanagement has been a significant factor in this degradation, since urban streambeds were typically used to remove waste products and many have been covered to lay sewer lines at these low elevations. Stormwater drainage systems pipe water rapidly away from cities instead of using it for vegetation and cooling evaporation. New studies on broad measures of public health reinforce the importance of water-related factors, such as air quality, bio-diversity, temperature control, recreational opportunities, and aesthetics of restored streams, trees, and other vegetation in cities and towns (Grant et al., 2010).

2.3.2 Enhancing the commons through smart, clean, and green design

Advocates of a 'green economy' also assert that governance reoriented to ecosystem protection and restoration, in particular greenhouse gas reductions, can be good for society (Inslee and Hendricks, 2008). Climate change can be slowed, while at the same time 'green jobs' are created, air pollution is lessened, and costs of energy are reduced. The 'human capital' of the new skilled workers in water and other sustainable resource management can lead to higher productivity in the economy and a rebuilding of middle-class incomes (Leonhardt, 2009). Advocates for a new paradigm in water management are beginning to make similar arguments.

There are some broad hints for how a comprehensive restructuring of infrastructure, institutions, and policy can serve society and ecosystems well. Traditionally, it has been thought that environmental protection is at the expense of social and economic wealth and consumption. But the new paradigm suggests that this is a false dichotomy. Because nature can be worked with in a smarter and cleaner way, ecosystem services can be restored in cities and many of the costs of municipal services reduced. More importantly, communities can be revitalized and the green jobs economy expanded.

These win-win opportunities stem from the core principles and practices of the new paradigm. An important opportunity is the immense productivity of natural systems that can be captured by society in ways that involve far less environmental degradation. The industrial model has been based on the linear model of mining or 'extracting' resources, using once, and dumping wastes. Current water systems use treated water for all purposes, when potable water use is a small fraction of overall need, and require substantial energy and chemicals for construction, pumping and treatment (Shrier et al., 2009). Wastewater collection and treatment is also energy and resource-intensive and, if effluent is released in rivers and ocean outfalls, can lead to depletion of groundwater and poisoning of the environment.

But natural systems have 'cleaner' resources that can be tapped with much less damage to the ecosystem. Constructed wetlands and on-site soils treatment are examples of low-cost and low-energy treatment options for small communities, which also replenish groundwater (Clement et al., 2010). Trees and rain gardens are effective alternatives to conventional drainage pipes and lagoons or underground storage tanks for stormwater management. Rainwater can be harvested and used on-site, thereby reducing the need for importing water through long-distance water lines.

Water demand can be reduced dramatically in homes and industry, both through water-efficient appliances or production processes, and more importantly, wastewater can be treated and reused for multiple non-potable purposes (Shannon, 2010). Water conservation and reuse can lower the energy and treatment costs of a municipal drinking water utility and the energy costs of running appliances, in particular hot water heaters, in homes. In a growing metropolitan area, these reductions in water demand can forestall the need for new extractive reservoir and water transport projects and new coal-fired or nuclear plants (Clerico and Kulik, 2011).

Wastes can be productively turned into resources as well. Conventional wastewater utilities generally release partially treated wastewater into surface waters and in some instances into groundwater, where dilution reduces concentrations further. But this used water and the embedded organics and chemicals can be seen as resources, not as waste (Shannon, 2010). Treated wastewater can be reused for non-potable purposes such as toilet flushing or landscape irrigation. Biogas digesters can recover energy from wastewater, while composting and urine-diverting toilets can recycle nutrients into fertilizer. Oyster farms can filter out and utilize excess nitrogen in coastal waters (Massachusetts Oyster Project, 2010).

Opportunities also exist in integrated design, rather than specialized thinking and practice. To paraphrase, the 'sum of the conventional parts' in the traditional approach has been much less than the 'whole' in infrastructure services. Integrated design can increase productivity of the larger system, while also serving the separate functional needs of the parts (Kirschenmann, 2005). For example, a recreation department and a stormwater utility can collaborate on the design of an urban park that includes trees and rain gardens. This park design can achieve multiple purposes of recreational space and stormwater retention and treatment at less cost than conventional drainage. Another way of putting this is that one agency's 'positive externalities' are another agency's core mission, and both agencies can benefit from collaboration.

Other resources to be tapped from nature are the models for efficiency and high-performance of its organisms and systems. Biologists and chemists are looking more and more to nature for lessons on how to re-engineer products and processes (Shannon et al., 2008). For example, permeable reactive barriers can intercept and treat groundwater, much as natural riparian zones do along rivers and estuaries (Lombardo, 2010). Rotating blades based on nature's spiral designs can increase water circulation in ponds and estuaries and promote natural water-cleansing (SolarBee, 2010). Kidneys filter wastes without high-energy demands and researchers are seeking to mimic these processes in wastewater treatment membranes (Wilderer, 2006).

In addition, as Ian McHarg wrote, by locating activities in the most appropriate places in a watershed, natural resource 'streams of value' can be tapped with less cost and disruption (McHarg, 1969). McHarg laid out guidelines for locating farms, ports, forests, wildlife corridors, and cities in places in the watershed that best matched the natural resource base or that avoid disrupting valuable ecosystems services within the watershed.

There are lessons to be learned, as well, from 'networks' of 'nodes' and 'links' in nature that increase resilience and adaptability to shocks and stresses in the system (Janssen et al., 2006). Hurricane Katrina and other natural disasters, along with projections of climate change instabilities, have highlighted the brittle quality of conventional water management approaches. A large centralized system is more vulnerable to damage from external shocks and equipment failures than a series of smaller, modular units, which are also put back in use much more quickly.

Examples of these transformative shifts where nature's design principles are being applied are now widely appearing in manufacturing, energy, and agriculture. As companies investigate manufacturing processes, they are discovering that efficiency improvements and capturing of resources instead of generation of waste can actually save money (Hawken et al., 1999). In the energy sector, energy-efficient appliances and clean sources of wind, solar, vegetative cooling, and building design can save money, reduce greenhouse gases, and generate jobs. In agriculture, complex plant and livestock rotation and harvesting can utilize solar energy and recycle nutrients, thereby avoiding expensive petroleum-based fertilizers, and producing healthier food with less runoff. Such farming practices actually increase the health of soils, rather than deplete them (Pollan, 2006).

Some of these innovations are about relearning intelligent ways to meet our needs by what we would ironically consider more 'primitive' societies. Older cities in the deserts, for example, rely on thick walls and street designs to capture desert winds to cool buildings down. Engineers and architects in Masdar in Saudi Arabia are trying to learn from these practices (CH2MHill, 2010).

Science can transfer these lessons into new technologies as well. New cities can install wind turbines and solar panels to capture and use energy in distributed locations. A forest moderates temperatures, through evaporative cooling and condensative warming, and a city can similarly use vegetation to lower heat in buildings and save energy. Rural villages in India have typically recycled nutrients from food to sewage and back onto the farmfields, and modern 'eco-sanitation' is seeking to replicate this nutrient-recycling process in developing and developed countries.

As of yet, these practices and concepts in water fall under the various categories of 'smart, clean, and green', 'integration', or 'biomimicry', resilience, etc., but there is no overarching design frame or 'unifying theory' that has emerged. One possible statement for water management could be: 'restore and replicate natural systems at the local scale. The benefits to the larger Commons will follow.' But, the densities of cities and the import and export of natural resources complicate this design principle. Another metaphor is the 'network', but the analogies for nodes and links need to be better identified. Another phrasing recently introduced is 'self-diagnosing, self-healing, and self-repairing systems' (Amin and Stringer, 2008). The value in fashioning a comprehensive design approach is in the identification of research and pilot project opportunities and in optimizing the benefits of each new building or infrastructure project that becomes available through new construction or repairs.

2.4 DESIGN PRACTICES AND VALUES IN THE NEW WATER PARADIGM

The internal design structure of a new water management paradigm consists both of a new set of technologies and practices and a new range of values generated by the approach. New 'smart, clean, and green' technologies and practices provide the means and opportunities for improvement. The increased value of services provides the rationale and motivation for heading down this new path. Some of these avoided costs or increased benefits can be captured directly by consumers, private companies, or communities, others help restore eco-systems and economies in the broader Water Commons.

In the first instance, in which a variety of actors or institutions can directly capture the monetary benefits of change, the new technologies and practices can either be incrementally cheaper than conventional approaches or provide additional quality, reliability, and services. Decentralized systems utilizing effluent pressure collection methods and soil-based dispersal or reuse, for example, are significantly less expensive than a centralized gravity sewer and large treatment plant in a rural community (Clement et al., 2010). Use of permeable barriers, oyster farming, or composting toilets can reduce costs for nutrient removal even further (Lombardo, 2010; Massachusetts Oyster Project, 2010). Water conservation and reuse technologies can allow a growing city to avoid expensive new water import or storage projects.

Another opportunity is in the capturing of the economic value embedded in pollutants or resources in the system, through methods of reuse and recovery. As described earlier, used water, energy, nutrients, and chemicals are all being looked at for productive reuse and capture and as potential revenue generators. Recent cost studies in British Columbia suggest that the market prices for these 'wastes' may be high enough to turn wastewater and energy infrastructure into a profitable enterprise, where the revenue from sale of recovered resources exceeds the cost of building the infrastructure in a new way (O'Riordan et al., 2011). Biogas digestors that recover energy for district heating or nearby industries, or sewer 'heat mining' are particularly valuable options.

A major economic opportunity is in the generation of secondary benefits, or positive externalities, as new water, wastewater, and stormwater infrastructure is operated. As earlier examples have suggested, green infrastructure for stormwater retention and treatment provides substantial other services in communities. Trees provide shade and lower air conditioning costs in surrounding buildings, remove pollutants from air, and bring nature back into cities. Studies show that students learn better, patients heal faster, and urban residents are generally happier and more intelligent when they are surrounded by trees, gardens, and parks. Expenditures can be reduced or performance enhanced for municipal service agencies, including education, health, recreation, and others. Numerous design, installation, and long-term maintenance jobs can be created as well, which improves the local economy and tax base.

Finally, there is economic value in resilience and risk-avoidance, which municipalities, insurance companies, and bond markets are beginning to recognize (Leurig, 2010). In the future, a sharp rise in costs of water and energy, or a catastrophic or steady increase in floods or droughts predicted from climate change, can be highly-disruptive to industry and municipalities. For example, a recent drought in the Southeast U.S. threatened Atlanta's domestic water supplies and the power plants that need to withdraw huge amounts of water for cooling. There is a rationale for spending more money now on reducing water consumption or increasing stormwater retention and groundwater replenishment in order to avoid future crises in public health and spending.

2.4.1 Capturing the benefits in practice

Examples of these cost-savings and value-added services in the new water paradigm are in green subdivisions or infill developments, schools, federal buildings, multi-family housing, and military bases, where property owners are attempting to either save money or create new value or both. In the early stages of 'low impact development', it began to be clear that developers could both save money on stormwater management, if done with natural systems infrastructure in a new subdivision and also garner higher home prices from buyers who liked the open space, recreational areas, wildlife habitat, etc. These factors have become a major tool in convincing developers to adopt 'green' practices and municipalities to adopt local ordinances that facilitate and encourage such measures.

Similarly, a few municipalities have begun to explore and understand the cost-savings and benefits of integrated, multi-benefit infrastructure. As suggested earlier, cities in British Columbia are starting to look at integrated water and energy planning, such as satellite wastewater treatment at pumping stations that can lower energy costs for the water utility and generate revenue from biogas recovery and water reuse. Los Angeles has calculated millions of dollars of savings from coordinating stormwater, solid waste, and drinking water around a rainwater harvesting approach (Lipkis, 2009). Seattle has started to experiment with long-term capital planning exercises that involve all relevant agencies (Antonoff et al., 2009).

These lowered costs, reuse and recovery values, and secondary and resilience benefits can all transform the economics and practices in water management, as private market participants and local communities recalculate their budgets and options. Much of the water and energy footprint of a city or town can be reduced as new design principles for water management are incorporated into new building and rebuilding projects. But negative externalities beyond the borders of cities and towns are likely to persist, even with these changes.

2.4.2 Watershed and planetary externalities

Within a watershed there can be dramatic impacts from water or wastewater practices of human settlements. Profligate use of surface water and groundwater has drained water from streams and from soils and sent water out eventually to oceans. High energy consumption has led to huge water withdrawals for nuclear and coal-fired power plant cooling. Toxics passed through wastewater treatment plants have led to changes in frogs, fish, and other wildlife, eutrophication from nutrient over-enrichment has destroyed lakes and estuaries, and cutting of vegetation has led to a warming and drying of land. Municipalities that implement smart clean green practices in infrastructure and in incentives for developers and industries for their own purposes, are still unlikely to take full account of these external impacts.

Recently, an international group of scientists has suggested that there are nine planetary boundaries or thresholds that we need to be careful not to exceed Climate change is the best-known of these, and water mis-management, through reductions in evaporative cooling and changes in atmospheric water cycles, are a factor in warming. But, water is implicated in other eco-functions, including biodiversity, nitrogen cycles, phosphorus cycles, ocean acidification, land use changes, and chemical pollution (Rockstrom et al., 2009).

The economics of these watershed and planetary externalities are an increasing focus of 'ecosystem services' studies, where the costs of depleting or benefits of maintaining or restoring natural water-related systems are calculated (USEPA, 2010). Relatively more progress has been at the watershed scale where, for example, the value of wetlands for water purification and fisheries production can be calculated. At the global level, the science of eco-functions is less certain and the monetary and non-monetary value in avoiding planetary thresholds is very large, but not well understood or documented.

2.5 POLICIES TO RESTORE THE WATER COMMONS

2.5.1 Current governance constraints

Current policies, institutions, and advocacy in water management are mired in conventional practice and thinking. Legislation in years past was directed at provision of clean water to cities (and farms) and at reducing exposure to contaminants in sewage in a manner known at the time, through long-distance transport of both. Government subsidies and regulations have perpetuated this design approach. Clean water advocates have largely accepted the notion that more funding and tighter enforcement would be the keys to improvement without considering the entire picture of the benefits of changing the basic approach.

The role of regulatory programs in perpetuating a conventional infrastructure paradigm is problematic. Generally, local developers are forced to deal separately with a wide range of conservation, health, building, planning, and other agency rules and permit requirements based on traditional practices and are allowed less flexibility in design than they would need to advance a 21st Century 'smart, clean, and green' agenda (Stebbins, 2005). Ordinances should be drafted to encourage, rather than stymie, the new agenda.

Municipalities can be greatly constrained by federal and state mandates in water quality, water supply, flood control, transportation, fire safety, endangered species, and other services. Separately, these mandates advance siloed agendas for the public interest, but collectively they may constrain the ability of communities to implement more holistic, sustainable approaches. This is one of the tensions in green infrastructure, for example, where enforcement actions have in the past perpetuated large underground storage tunnels to deal with combined sewer overflows, rather than distributed retention in green roofs, swales, and tree-plantings. Fire codes have also perpetuated wide streets and cul-de-sacs, and sprawl development.

A series of water crises and science-based design opportunities are challenging this system, but the tendency is still to think incrementally and cautiously. With some flexibility in regulatory programs and some support for innovation, scattered pilot projects have been built by developers or municipalities in the U.S. These are not without value. Demonstrated success (or failure) with new technologies and designs can expand the knowledge base and lower the risks of reform.

But a new paradigm in water does require a fundamental rethinking of governance and civic activism that matches in scale and scope the radical shift needed in technologies and practices. Reference points for a new policy and civic agenda are in the larger conversations about the failings of markets and the need for a restored social and economic Commons described earlier. New policies and advocacy for sustainable energy and agriculture also offer important analogies.

2.5.2 New pricing principles and incentives

The 'Tragedy of the Commons', as discussed earlier, is a metaphor for incorrect pricing or lack of cooperative management of natural resources and the resulting destruction of ecosystems. Current low prices for water and wastewater services encourage the wasting of increasingly scarce water resources. Failure to charge for the externalities of eco-system damage similarly encourages disruption of water hydrologies and increased releases of pollutants.

Higher pricing of water is now being recommended by mainstream economists and by corporations that rely on large and stable water supplies, including power plants, bottling companies, computer chip manufacturing, and others (McKinsey, 2009). The argument is made that full-cost pricing would spur more efficient water allocations and would encourage a search for and adoption of water-efficient designs, fit-for-purpose treatment, and reuse and reclamation. Resistance to such 'commodification' of water comes from social justice groups who argue that access to water should be a human right and that the poor should not be priced out of the water market by corporations or customers that have the ability to absorb higher rates (Barlow, 2010).

More importantly, higher prices in the current utility management structure would not necessarily lead to a shift to more sustainable practices. The entrenched traditional business model for municipal 'monopolies' would likely be perpetuated through the construction of large desalination plants, longer-distance pumping and piping of water supplies, more advanced wastewater treatment plants, and purchasing of water rights from agriculture and other sectors, the costs of which would now be covered by higher water and sewer rates.

Instead of a blunt-force rise in water and sewer rates, a more complex change in incentives and regulations is required that matches the complexity of new participants, benefits, and externalities in the new water paradigm. Subsidies, regulations and civic advocacy all would need to be implemented to influence more directly the behaviour of new market participants in the decentralized and integrated approach, either by limiting projects or designs that collectively threaten the environment (negative externalities) or promoting projects that collectively enhance the environment (positive externalities).

Improvements should also be internalized at the lowest scale possible, a principle termed 'subsidiarity'. In water management, there are the following scales: building, subdivision or neighborhood, municipality, watershed, and global ecosystem. Generally, at each scale, market participants may initially need some encouragement or information to adopt practices or technologies that are in their interest. A heavier hand of pricing or mandates will be required to deal with externalities that they impose collectively at higher scales.

2.5.3 Learning from other sectors

A key insight from the energy sector is that market participants include far more than energy utilities and multi-national corporations in coal, oil and natural gas. Homeowners and other property-owners become much more active customers for energy-efficient appliances, solar panels, etc. Scientists and venture capitalists develop wind, solar, and biogas technologies at a variety of scales. Architects incorporate LEED and other Green Building standards in their plans and drawings. Planners and a range of municipal agencies and civic organizations search for more efficient and renewable energy designs and strategies.

The energy sector has made substantial progress, though not yet complete, in defining a new role of government and civic commitment in addressing greenhouse gas emissions, that includes education campaigns, weatherization grants for low-income homeowners, vehicle mileage requirements, tax credits and preferential lending for green projects, funding of research and pilot projects, early adoption in federal facilities,and other strategies. The agricultural sector has also developed local food campaigns, pilot projects in natural system farming and urban gardens, incentives for conservation and nutrient-reduction approaches, and others.

Similarly, in water management, embedded reuse and green infrastructure 'nodes' in homes, subdivisions, and commercial establishments engage a wide range of private firms, non-profit groups, and other city agencies (such as parks and recreation, housing, job training, etc.), homeowners can install composting toilets or green roofs, and the developer and property-owner will have many more choices for technologies, and for design and ongoing maintenance services. Municipalities will have more complex and highly-productive new roles in coordinating municipal utilities and agencies internally and in overseeing the new private and non-profit sector externally, through ordinances, incentives, education, and inspections.

New water policies could include: rebates for water-saving appliances; higher water rates to induce conservation and reuse; subsidies targeted at low-income ratepayers to install new technologies; mandates for onsite stormwater management practices and stream restoration; integrated full-cost assessments for federal grants and loans; water banks, where nonprofits pay for reductions in water use; development of ecosystem service markets, modeled on carbon trading markets; support for cleantech investments; funding of academic research and pilot projects; and other institutional changes, such as new leasing systems where the developer can capture the longer-term benefits of reduced water use (and not the tenant).

Other private and non-profit or community groups could also provide incentives for implementation of smart, clean, green strategies. Investors and insurers are beginning to educate industry on the benefits of sustainable water management in surrounding communities, in terms of both guaranteed supplies and public image. Borrowing costs will increase for utilities that lack plans to deal with climate change and other instabilities. NGOs can work to shift the awareness and ethics surrounding water, so that homeowners, churches, schools, etc. choose designs and practices that increase environmental stewardship and address the human right to water.

2.5.4 Achieving an optimal mix is difficult…

The design of an optimal mix of government and other strategies will depend on calculations of the responsiveness, or 'elasticity' of demand or supply, of various market players and communities to incentives and opportunities. There are types of 'market failure' to keep in mind as well, where markets will underperform because individual and societal benefits and costs are misaligned. For example, participants in the market may lack necessary information, under-invest in public research, unfairly usurp resources from others, exaggerate the risks of innovation, or otherwise act irrationally.

Many of the incentives and information campaigns might be considered 'transitional'. Over time, pilot projects can demonstrate the reliability of new approaches, good models will have been developed for municipal planning, the costs of technologies will fall as production increases, and in general, the benefits of the new water management paradigm will be more widely understood and internalized by homeowners, developers, companies, and municipalities. The need for special government or civic education and activism will be reduced, at least to some degree, as citizens widely understand the benefits of smart clean green practices in their own lives and in their communities.

However, there will continue to be a need for government to address the broader externalities in the watershed and global Commons. The watershed-scale perspective is an important one for providing incentives or imposing penalties and mandates on market participants beyond what they might otherwise do in their pocketbook self-interest. The financial costs or penalties of extracting water or dumping wastewater need to be raised or the financial benefits of conservation and waste recovery increased to reflect the ecosystem services that will be lost or restored, or both. Pilot projects for better management need to be developed and stronger watershed authorities that collaborate with municipalities must be created.

In some watersheds, the primary concern may be water supply, in others water quality, but in all watersheds will be concern for bio-diversity, wetlands protection, cooling from vegetation, nutrient cycles, and toxics. Increasing droughts and heavy rainfall events will intensify these challenges. And, as McHarg has suggested, urban planning and infrastructure should be combined with agriculture, forestry, energy and other resource management, where there is really one water system to be concerned about, with interdependent activities. Integrated resource management at the watershed-scale will be required to deal with these crises in the Commons.

International institutions also need to operate in a longer time-frame than markets or municipalities typically do, and calculate costs and benefits of actions or inaction with a lower social discount rate. A primary example of this is phosphorous depletion. It is now understood that supplies of phosphorus, which is necessary for all plant growth, will run out in about fifty years (Langcuster, 2010). Water is the transport mechanism for phosphorus to move off farm fields and, via food consumption, through sewer systems and out into lakes and oceans. Private market prices do not yet reflect the shortages that will occur when phosphorus mines are empty, because these events are far in the future. But society has a much greater concern and a lower interest rate for future events, and incentives for phosphorus recovery or penalties for continued phosphorus pollution, which also causes eutrophication of lakes, should be implemented soon.

Much of the political economy field has feared that these complex changes in the way we manage and govern water are too difficult to achieve, suggesting instead that we can continue to look for gains in the traditional industrial model, by working on conservation measures with a few large utilities, corporations, and farms that we can sit down and negotiate with (McKinsey, 2009), and by putting off protection of the Commons until the costs of inaction elicit a crisis response (Schor, 2010). But these models assume that there are new supplies or substitutes for a resource in short supply, and this is not the case for water. As historians have shown, this inattention to damage in the water Commons has ultimately led to the collapse of ecosystems and societies, from which there is no turning back (Hedges, 2011).

Emerging corporate thinking is that the looming water crisis is largely about supply and is primarily a local concern, but that view is dangerous and wrong. The interdependencies of ecosystems, the economy, and society as mediated through water at a global scale are profound. Loss of bio-diversity, and evaporative cooling cycles matter to everyone, wherever they take place. Virtual water markets exist, where water is embedded in food and manufactured products that are shipped all over the world. Rich countries may be able to purchase a high share of this global water market, and some countries have adequate rainfall. Nevertheless, all countries and international institutions should consider a shift to smart clean green water management that protects and restores healthy water systems as a means both to avoid global ecosystem collapse and to assure the human right to water and sanitation everywhere. These measures also help to prevent water-related wars and conflicts as a matter of national and international security.

2.5.5 … and urgent so we must act artfully now

Developing and implementing a new water paradigm thus requires significant interdisciplinary collaboration and participation by civil society. Governments, foundations, and a wide variety of professional and advocacy groups need to create substantial spaces for conversation and invention. Markets can experience the 'creative winds of destruction' as new products drive out the old. But, in water management, there are complex and intricate relationships among private companies, agencies, civil society, and the public that open or close markets, by virtue of standards, permits, etc. These entities must together reshape and define the boundaries and incentives of markets to allow for these innovations to emerge. Citizens in local communities also must lead their municipal officials to set high standards for sustainability and to invest resources in pilot projects and inter-agency collaboration.

An 'artful' new policy frame will seek to maximize the strengths of markets, but direct those markets toward protection and restoration of the Water Commons, rather than 'commodify' water. Current policies protect public health in important ways, but also thwart the discovery of efficiencies and innovative technologies and designs. Market forces do need to be unleashed, but only if goals, incentives, and safeguards are in place to advance the public interest, including the health and functioning of ecosystems and communities worldwide.

Ecological crises in the Commons are too great and the benefits of a smart clean green approach are too promising to turn away from the challenges of complexity. Participation in smart clean green approaches by homeowners, developers, start-up companies, and institutions like schools, colleges, and churches, is all possible. Inter-agency collaboration is difficult at the municipal level, but achievable with civic leadership. Widely-shared responsibilities and aspirations have guided many sustainable indigenous communities in managing their Common resources in the past, and community-based management, markets, and ethics can be restored in 21st Century ways in modern cities and towns.

2.6 CONCLUDING REMARKS

A water paradigm shift requires an affirmative government and civil society vision and commitment for change, which will only emerge from a substantial collaborative dialogue. Nature's patterns can be restored in cities to provide water and energy services. Markets and agency missions can be adjusted to advance lighter urban footprints and healthier communities. And, the Water Commons and civil society can be revitalized in the rebuilding of the 'Common Wealth'.

The touchstone for the new paradigm is in the phrase 'water is at the heart of all life.' Inevitably, such concepts as 'cooperation', 'interconnection', 'resilience', and 'abundance' will emerge. Unlocking the correct metaphors and design principles for water's work in the Commons is an important shared responsibility for urban water managers of the future.

REFERENCES

Abrams, R. and Hall, N. (2010). Framing Water Policy in a Carbon Affected and Carbon Constrained Environment. Natural Resources Journal, Vol. 50. (Winter 2010).

Antonoff, J. et al. (2009). Sustainable Infrastructure Initiative: Interdepartmental Planning for Better Capital Projects. City of Seattle Department of Planning and Development, Seattle, WA, USA.

Amin, S.M. and Stringer, N. (2008). The electric power grid: Today and tomorrow. MRS Bullet. 33. http://www.mrs.org/bulletin (accessed 12 December 2008).

Barlow, M. (2010). Our Water Commons: Toward a new freshwater narrative. The Council of Canadians, West Ottawa, ON, CN.

Braungart, M. et al. (2008). Ecological Design and Engineering for Urban Environments. Industrial Technology, Vol. 4(3), (September 1, 2008).

CH2MHill. (2010). Building a Carbon-neutral City: Masdar City, Abu Chabi, UAE. Boulder, CO.

Clement, T. et al. (2010). Sustainable Water Resource Management Volume 3: Case Studies in New Water Paradigm. EPRI, Palo Alto, CA.

Clerico, E. and Kulik, D. (2011). Natural Systems Utilities: A Sustainable Water Company. Presentation to Water Infrastructure Finance Commission, Boston, MA, USA.

Cooper, M. (2009). Aging of Water Mains is Becoming Hard to Ignore, New York Times (April 17, 2009).

Diamond, J. (2005). Collapse: How Societies Choose to Fail or Survive. Penguin Books, London.

Fraker, H. and Wurster, W. (2008). Sustainable Neighborhood 'Eco-blocks' in China. http://bie.berkeley.edu/ecoblocks (accessed 12 April 2009).

Grant, J. et al. (2010). The Value of Green Infrastructure: A Guide to Recognizing Its Economic, Environmental, and Social Benefits, Center for Neighborhood Technology, Chicago, IL.

Hardin, G. (1968). The Tragedy of the Commons. Science, 162(3859), 1243–1248.

Hawken, P., Lovins, A. and Lovins, L.H. (1999). Natural Capitalism: Creating the Next Industrial Revolution. Little, Brown and Company, Boston, USA.

Hedges, C. (2011). This Time We're Taking the Whole Planet With Us, CommonDreams.org (March 7, 2011).

Hoegh-Guldber, O. et al. (2007). Coral Reefs Under Rapid Climate Change and Ocean Acidification. Science, Vol. 318, No. 5857. (December 14, 2007).

Inslee, J. and Hendricks, B. (2008). Apollo's Fire: Igniting America's Clean Energy Economy. Island Press, Washington, D.C.

Janssen, M. et al. (2006). Toward a Network Perspective of the Study of Resilience in Social-Ecological Systems. Ecology and Society, Vol. 11, No. 1.

Kirschenmann, F. (2005). Spirituality in Agriculture. Concord School of Philosophy, Concord, MA. (October 8, 2005).

Kravcik, M., Pokorny, J., Kohutiar, J., Kovac, M. and Toth, E. (2008). Water for the Recovery of the Climate: A New Water Paradigm. Typopess-publishing house, Kosice, Slovakia.

Langcuster, J. (2010). Experts fear critical phosphorus shortage. Western Farm Press. (October 19, 2010).

Leonhardt, D. (2009). The Big Fix – Can Barack Obama Really Fix the U.S. Economy. The New York Times. (January 27, 2009).

Leurig, S. (2010). The Ripple Effect: Water Risk in the Municipal Bond Market. CERES, Boston, MA.

Lipkis, A. (2009). Integrating Resource Planning: Investing in Smart Green Infrastructure for Sustainable and Healthy Cities. Presentation at Water Environment Research Foundation Briefing: Smart Clean Green Water Management, Alexandria, VA. (February 25, 2009).

Lombardo, P. (2010). Representative Groundwater and Stormwater Project Descriptions. Lombardo Associates, Newton, MA.

Massachusetts Oyster Project for Clean Water Removal from Wastewater. (2010). Exploration of the Cost-Effectiveness of Using Oysters for Nitrogen Relative to the Cost of Constructing Incremental Wastewater Treatment Plant Capacity. Boston, MA.

McHarg, I. (1969). Design With Nature. Natural History Press, The Falcon Press, Philadelphia, USA.

McKinsey & Co. (2009). Charting Our Water Future: Economic Frameworks to Inform Decision-making. Water Resources Group, London.

Morrison, J. (2010). Guide to Responsible Business Engagement with Water Policy. Pacific Institute. The CEO Water Mandate, Pacific Institute. Berkeley, CA, USA.

NRC (National Research Council). (2009). Sustainable Critical Infrastructure Systems – A Framework for Meeting 21st Century Imperatives. The National Academies Press. Washington, D.C.

Nelson, V. (2007). The Baltimore Charter for Sustainable Water Systems. http://www.sustainablewaterforum.org (accessed 11 April 2009).

O'Riordan, J. et al. (2011). Integrated Resource Recovery Study: Metro Vancouver North Shore Communities. Fidelis Resource Group, Victoria, B.C., CN.

Pollan, M. (2006). The Omnivore's Dilemma: A Natural History of Four Meals. Penguin Books, New York.

Rockstrom, J. et al. (2009). A Safe Operating Space for Humanity. Nautre (September, 2009).

Schor, J. (2010). Plenitude: The New Economics of True Wealth.The Penguin Press, New York.

Shannon, M., Bohn, P., Elimelech, M., Georgiadis, J., Marinas B. and Mayes, A. (2008) Science and technology for water purification in the coming decades. Nature, 452(20 March), 301–310.

Shannon, M. (2010). Energy Recovery from Waste Water. ARPA-E Pre-Summit Workshop, VA, (March 1, 2010).

Shrier, C. et al. (2009). Integrated Water-Energy Planning. The Water Report, Issue #66. (August 15, 2009).

SolarBee. (2010). Long-Distance Circulation and Mixing: An ecological approach to water quality management. Dickinson, ND.

Stebbins, J. (2005). Building Blocks of Sustainable Development. Project Design Consultants, San Diego, CA.

U.S. EPA. (2010). Ecosystem Services Study Focuses on Albermarle-Pamlico Watershed in Carolinas. Ecosystem Services Research Program, Washington, D.C.

Walker, B., and Salt, D. (2006). Resilience Thinking: Sustaining Ecosystems and People in a Changing World. Island Press, Washington, D.C.

Wilderer, P. (2006). Membrane Technology Revolutionizes Water Treatment. International Water Association Small Water and Wastewater Systems Conference, Mexico City, Mexico.

World Business Council for Sustainable Development. (2010). Vision 2050: The new agenda for business. WBCSD, Conches-Geneva, Switzerland.

Chapter 3

Transitioning to the water sensitive city: the socio-technical challenge

Rebekah Brown

Monash University, Centre for Water Sensitive Cities, Melbourne, Australia

ABSTRACT

This chapter begins with the premise that despite the many benefits the traditional centralised urban water servicing model has provided for society, it is increasingly ill-equipped to cope with the prevalence of extreme weather events, climate uncertainty and other variable socio-technical trends. A new model of adaptive water management seeks essential shifts in management approaches; broadening system boundaries, examining underpinning philosophies, reassessing expertise and skills required, reviewing service delivery goals, reconsidering the role of the public and re-evaluating risk management strategies. Based on extensive socio-institutional research of Australian experiences managing urban water under pressures such as modern environmentalism and extreme and prolonged water scarcity, this chapter presents an heuristic tool for informing and strategising these significant shifts needed to transition to more sustainable urban water management regimes. The chapter plants a future vision of Water Sensitive Cities as a moving target, identifies current barriers preventing moves toward this future, and provides some strategies employed to overcome such barriers drawn from empirical research on Australian cities. At its core, the chapter provides a commentary on the current state of urban water reform and posits ideas to stimulate the future research agenda on urban water paradigm transitions.

Keywords: adaptive water management; transitions strategies; water sensitive cities

3.1 INTRODUCTION

Cities across the globe are facing increasing challenges for managing water. Water is critical for the health, viability and development of cities; yet urban population growth and climate variability are placing pressure on resource availability and already stressed ecosystems. At the same time, water services infrastructures are reaching the end of their lifespan in developed countries, while developing countries race to meet growing needs, often importing traditional models and standards from developed nations, arguably unsuited to their conditions and socio-political contexts. These prevailing water management technologies, modelled on large centralised potable supply systems, do not always offer urban communities the flexibility needed for meeting sustainable development goals, nor the ability to address future conditions. It is also increasingly recognised that, along with changing consumption habits and expectations, sustainable development is more likely to be achieved through a diverse suite of alternative supplies, such as recycled wastewater, greywater, stormwater and decentralised technologies. Utilising an assortment of water supplies augments centralised infrastructure, while protecting waterway health, and therefore builds flexibility into servicing options. Despite policy rhetoric, proven technology options and well performing demonstrations projects, modern cities have had limited success at implementing and managing these complex water supply and waterway health protection practices in a cohesive and mainstream way.

3.2 TRADITIONAL & ADAPTIVE MANAGEMENT

It is well recognised that urban systems are designed on the basis of prediction and control (Milly 2008: Newman 2001). Modelling on past data provides water managers with a fixed storage and delivery capacity to build into the system. Even with provision for some of the largest per head storage capacity in the world, this conventional approach failed to predict and cope with recent extreme and prolonged dry conditions experienced in southern Australia. The traditional approach, with its emphasis on reducing technical uncertainty (Newman 2001) in the face of unprecedented conditions, highly variable trends and unknown impacts on water availability, is proving to be less than effective. As Milly and colleagues (2008) sharply argue, the water services industry operates on a fundamentally incorrect assumption of stationarity which simplifies the urban water problem, thus privileging technocratic solutions (Pahl-Wostl *et al.,* 2009). Such a paradigm of efficiency through centralisation and optimisation, based on the assumption of predicable future conditions, has been deemed a 'mal-adaptation' response (Barnett and O'Neill 2010; Dawson 2007), in that it produces ongoing investment in approaches designed for permanence and supported by solely large centralised solutions. This is despite the existence of a host of technologies and management options aimed at improving overall system resilience, developed over the last two decades (Gleik 2003; Mitchell 2006; Wong 2006; The Barton Group 2005; Maksimovic and Tejada-Guibert 2001).

The development of these alternative approaches follows the consensus amongst water management scientists that to service urban communities and water environments under uncertain futures, building resilience through system integration and adaptability is required. Adaptive Water Management (AWM) is seen as the new underlying paradigm for planning, designing and operating water services. AWM starts from the perspective that urban water systems are complex systems where dependencies between the human-technological-environment sub-systems exist. Through interplay between these dependencies, functional properties and system behaviour and performance emerge, and therefore make uncertainty irreducible (Pahl-Wostl *et al.*, 2009). Given this basis, AWM seeks instead to recognise uncertainty and deal with its presence by increasing the adaptive capacity of the management regime. In doing so, uncertainty becomes an explicit driver of not only coping with change, but learning to do things better. This is in contrast with the traditional management regime, which is geared toward finding and developing largely fixed and inflexible solutions. Therefore the aim of AWM is to create urban water systems that are resilient to shocks and disruptions such as those caused by droughts, floods, and heatwaves and increase the opportunities for system adaptation, flexibility and transformability. Table 3.1

Table 3.1 Attributes of Traditional and Adaptive Urban Water Management Regimes (reproduced from Keath and Brown 2009)

Attributes	Traditional Regime	Adaptive Regime
System Boundary	Water supply, sewerage and flood control for economic and population growth and public health protection	Multiple purposes for water considered over long-term timeframes including waterway health, transport, recreation/amenity, micro-climate, energy etc.
Management Approach	Compartmentalisation and optimisation of single components of the water cycle	Adaptive, integrated, sustainable management of the total water cycle (including land-use)
Expertise	Narrow technical and economic focussed disciplines	Interdisciplinary, multi-stakeholder learning across social, technical, economic, design, ecological spheres
Service delivery	Centralised, linear and predominantly technologically and economically based	Diverse, flexible solutions at multiple scales via a suite of approaches (technical, social, economic, ecological etc.)
Role of public	Water managed by government on behalf of communities	Co-management of water between government, business and communities
Risk	Risk regulated and controlled by government	Risk shared and diversified via private and public instruments

compares the idealised differences between the traditional approach to urban water management and a more sustainable, adaptive regime.

The adaptive regime attributes highlighted above point to the need for integrated management approaches, and poses urban water governance arrangements significantly at odds with the traditional approach. This critique is based on the view that sub-optimal outcomes have been produced from the traditional compartmentalisation of water supply, sewerage and stormwater services. This compartmentalisation has been physical in terms of infrastructure design and operation, and institutional in terms of responsibility for service provision, operation and maintenance. Over time this has led to philosophical compartmentalisation and shaped perceptions of system boundaries that underpin current management arrangements and practices (Brown 2008; Brandes and Kriwoken 2006; Ashley 2005). Thus multiple commentators believe that the existing management regime poses significant barriers to change. They argue that rigid regulatory and other governmental mechanisms reinforce the compartmentalisation of infrastructure and service provision, leaving the sector ill-equipped for responding and adapting to complex sustainability challenges (Brandes and Kriwoken 2006; Wong 2006; Marsalek *et al.,* 2001; Newman 2001). These commentators thus argue that current progress toward more sustainable urban water management is too slow (Brown *et al.*, 2007; Harding 2006).

Adaptive urban water management regimes would emphasise a systems approach whereby interconnections between the management of each of the water streams (and other related functions such as land use planning) would deliver and protect multiple benefits. They would also be adaptive and ready to respond to unanticipated outcomes by being prepared for multiple potential future conditions. Therefore, investing in a level of strategic redundancy would be part of a resilient system. Such an approach is somewhat at odds with traditional urban water management whereby the most likely future condition (i.e. water scarcity) is often optimised; leaving systems potentially inefficient and vulnerable to other inevitable futures (i.e. water abundance) (Pahl-Wostl 2007).

3.3 ENVISAGING WATER SENSITIVE CITIES

While concepts such as integrated urban water management and adaptive water management offer alternative philosophical approaches to the traditional urban water management paradigm, urban water strategists still lack a clear vision or goal for the attributes of a sustainable water city. In 2007, nineteen Australian scholars representing seven disciplines participated in an envisaging process focussed on water and cities of the future. Ensuring socio-technical resiliency, and overcoming system (or city-wide) vulnerability to climate change and population growth, was considered an important starting condition for the Water Sensitive City. Building resilience in cities was recognised as complex and involves multiple stakeholders and disciplines that input into the process of water management in more deliberative and shared ways. They further concluded that future Water Sensitive Cities would ensure environmental repair and protection, supply security, public health and economic sustainability, through water sensitive urban design (WSUD), enlightened social and institutional capital, and diverse and sustainable technology choices (Brown *et al.*, 2007).

While there is not one example in the world of a Water Sensitive City, there appears to be cities that lead on distinct and varying attributes of the water sensitive approach such as, wastewater recycling, stormwater harvesting or end-use conservation practices. To further develop this thinking Wong and Brown (2011), presented a framework for envisioning the attributes of Water Sensitive Cities in different bio-physical and socio-political contexts. They propose three guiding principles or 'pillars' that would need to be integrated into the urban environment through urban design and planning. These include:

1. *Cities as Water Supply Catchments*: Cities would have access to a diversity of water sources in addition to the established convention of capturing rainfall-runoff from rural and forested catchments. These alternative water sources for cities could include a mix of groundwater, urban stormwater (catchment run-off), rainwater (roof run-off), recycled wastewater and desalinated water. These sources would be delivered through an integrated mix of centralised and decentralised infrastructure. Such a strategy of diverse water sources delivered at a mix of water infrastructure scales is to allow cities the flexibility to access a portfolio of sources at least environmental, social and economic costs. Each of the alternative water sources will have an individual reliability, environmental risk and cost profile. This pillar ensures that cities contain both centralised and decentralised water supply schemes, such as a simple domestic rainwater tank for non-potable use, through to city-scale indirect potable reuse schemes and a pipeline grid linking regional reservoirs.
2. *Cities Providing Ecosystem Services*: Cities water services would focus on providing low carbon urban ecosystem services (such a water quality treatment and city cooling) rather than degrading such services. New ecological landscapes incorporating green infrastructures can be achieved through the integration of urban landscape design with sustainable urban water management to buffer the impact of climate change (in particular increased frequency of extreme storm events) and increasing urban densities on natural aquatic environments and preserve, and/or re-establish, ecosystem services. Landscapes are the product of varying natural and human-induced forces, interacting within a regional and global ecosystem. Traditionally spaces in the public domain are essential features of public amenities, however in water sensitive cities these urban landscapes must be ecologically functional as well as providing for spatial amenity. Our urban landscapes need to be designed and retrofitted to capture the essence of

sustainable water management, micro-climate influences, facilitation of carbon sinks and potential use for urban food production.

3. *Cities Comprising Water Sensitive Communities and Institutions*: The capacity of institutions themselves to advance sustainable urban water management is essential. Unless new technologies are embedded into the local institutional and social context, their development in isolation will not be enough to ensure their successful implementation in practice. The social and institutional capital inherent in the city is reflected in (i) the community living an ecologically sustainable lifestyle and cognisant of the ongoing balance and tension between consumption and conservation of the city's natural capital; (ii) the industry and professional capacity to innovate and adapt as reflective practitioners in city building; and (iii) government policies that facilitate the ongoing adaptive evolution of the water sensitive city.

Therefore the approach of the Water Sensitive City would be underpinned by a flexible institutional regime and co-existing and diverse infrastructure. While such socio-technical perspectives have been identified as the most promising for addressing the need for resiliency and advancing sustainable development (Meadowcroft 2005), they have yet to be systematically applied to the urban water environment. Therefore the key question of 'how' to transform cities (by connecting best thinking and practice in urban water management, urban design, and social and institutional systems) into Water Sensitive Cities remains. The next section is a starting point for considering the question of how cities can transition to the Water Sensitive City.

3.4 TRANSITIONING TO THE WATER SENSITIVE CITY

The lack of a socio-technical heuristic tool for assessing and informing the transition of cities towards more sustainable practices has been a critical barrier for conceptualising significant changes over time. In an attempt to address this crucial gap, an *urban water transitions framework* has been developed through an historical and futures analysis of the changing institutional and technological arrangements supporting Australian cities urban water management practices over the last 200 years (see Brown *et al.*, 2009). The framework recognises temporal, ideological and technological contexts of different management paradigms, and is sensitive to other influencing contextual variables such as histories, ecologies, geographies and socio-political dynamics. This mix of contextual features has been described as the 'hydro-social contract' and encompasses dominant values or implicit agreements between communities, governments and business on how water should be managed (Lundqvist *et al.*, 2001).

The transitions framework (Figure 3.1) presents a typology of six types of dominant water management regimes (water supply, sewered, drained,

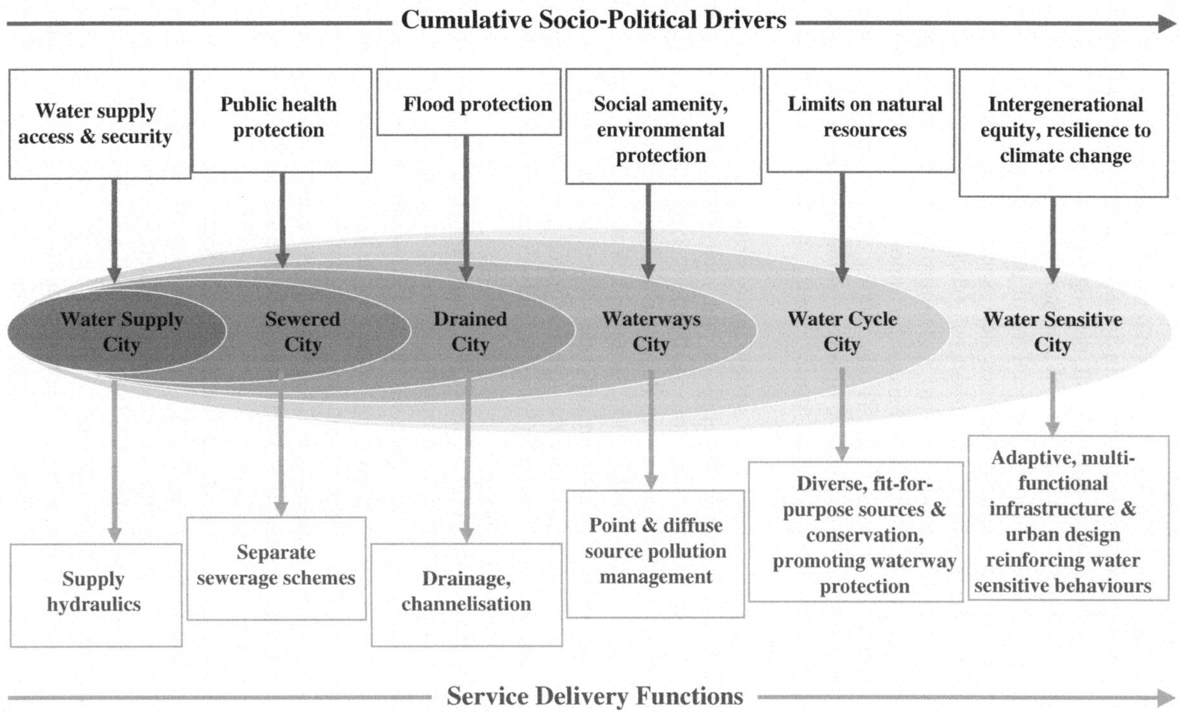

Figure 3.1 Urban Water Management Transitions Framework (Brown *et al.*, 2009)

waterways, water cycle and water sensitive) that represent a nested continuum of socio-political drivers and service delivery responses. This typology is based on the attributes of past and present hydro-social contracts in Australian cities and proposes potential future hydro-social contracts by anticipating social and institutional factors underpinning the principles of integration and resilience espoused in proposed future management paradigms such as AWM. The idea being that as cities progress towards the Water Sensitive City they accommodate additional and sometimes competing objectives of previous management regimes and therefore water management becomes necessarily more complex, but also more resilient to major system 'disturbances' (such as floods, droughts, heat waves and waterway health degradation) and improves its adaptive capacity to create opportunities from these disturbances for innovation and development or even the pursuit of new trajectories (Folke 2006). Therefore earlier types of management regimes are representative of more 'vulnerable' systems, when even small disturbances, such as extended storm events, are likely to cause dramatic social consequences.

Water Sensitive Cities encapsulate sustainability principles such as integrating social/environmental/economic imperatives and inter/intra generational equity. They represent a system that is more resilient, has the adaptive capacity to create opportunities from major system disturbances for innovation and even the pursuit of new trajectories (Folke 2006; Smit and Wandel 2006). Thus through socio-institutional drivers which incorporate sustainability principles, Water Sensitive Cities create sustainability through integrated environmental protection, diversity of resource sources and technologies, and new forms of environmentally and social conscious urban design.

Research by Brown and colleagues (2009) has shown that many Australian cities are currently progressing from the more conventional water supply, sewered and drained city management regimes towards the waterways city, with far more attention on protecting waterway health and addressing urban stormwater quality issues. Over recent years there has been a shift in Australian cities advocating more operational policies towards the water cycle city but this is yet to be realised as mainstream practice beyond the role of some progressive demonstration projects in practice.

The application of the transitions framework to other cities around the globe has suggested that many of the world's cities are located on the left half of the continuum (Duffy 2009), with frontrunners located somewhere between the 'Waterways' and 'Water Cycle' city. Reflecting on the commentary thus far, it could be argued that there appears to be a deep barrier to making the transition from the 'Drained City', representing a state of large technical, single source water supply systems, energy intensive reticulation and treatment systems and drainage systems that continually degrade aquatic ecosystems, to the more complex, adaptive practices represented on the right side of the continuum.

This is perhaps because the structure of the hydro-social contract is significantly challenged with the advent of the Waterways City, with the players in this contract expanding and responding to environmental protection and more social amenity needs. With the Water Cycle City the players and the distribution of functions and responsibilities need to be fundamentally re-structured compared to the comparatively more incremental structural adjustments associated with the transition from a Sewered to a Drained City.

The above commentary illustrates that the establishment of Water Sensitive Cities will require major socio-technical overhauls of conventional approaches, which is underpinned by fundamental transition(s) to the socio-institutional setting and particularly the hydro-social contract of urban water management.

3.5 BARRIERS AND OPPORTUNITIES FOR TRANSITIONING

The above discussion supports the current scholarly argument that change is slow due to the paradigm shift required, and the inability of current management regimes to begin to negotiate adjustments required in regulatory and other governance mechanisms to bring about socio-technical change in the physical water management system and its operational frameworks. Therefore the impediments to progress do not seem to always arise from the unavailability of technological solutions and scientific knowledge, but rather the failure to manage the social and institutional change processes that would enable the implementation of techno-scientific solutions in practice.

It has been suggested that the greatest challenge to enabling a significant transition in practice is the issue of path dependence (Geels 2004; Walker 2000). This phenomenon prevents the adoption of better alternatives when they are available, as historical investment in past practices continues due to the increasing returns provided by legacy infrastructure (Arthur 1989). Further work to examine the effect of path dependence on attempting AWM (Brown and Farrelly 2009) revealed a common and overarching suite of barriers including;

- insufficient skills and knowledge
- organisational resistance
- lack of political will
- limited regulatory incentives
- unsuitable institutional arrangements
- lack of community engagement
- policy failure at implementation

These barriers reveal a lack of suitable management capacity to enable a paradigm shift at both the individual and institutional level, which is consistent with other findings in the literature (Brandes and Kirwoken 2006; Harding 2006; Mitchell 2005; Saleth and Dinar 2005; Blackhurst *et al.*, 2004).

Therefore institutional reform for adaptive management, and the transitioning ingredients and/or processes for moving towards the Water Sensitive City remain elusive. Like most reform agendas, it is likely to require the consideration of options that are not immediately clear, technically or otherwise. Therefore it is very important to see more empirical studies of reform successes rather than just the barriers to change. In one such study, Brown and Clarke (2007) analysed the historical and socio-technical drivers across Melbourne, Australia, in its transition from a Drained to a Waterways city over the last 40 years or so. They found that the transition dynamics represented a complex interplay between a small network of issue champions (or change agents) and a suite of enabling context variables, important for mainstreaming the changes made, as listed below:

1. **Socio-Political Capital**
   Community, Media and Political
2. **Champions**
   Vision
   Multi-sectoral network
3. **Accountability**
   Coordination Processes
   Water Cycle
   Land-use Planning
4. **Reliable & Trusted Science**
   Academic leadership
   Technology Development
5. **Market Receptivity**
   Business Case for Change
6. **Bridging Organisations**
   Facilitates Science – Policy
   Facilitates Capacity Building
7. **Binding Targets**
   Measurable System Target
   Science, Policy and Development
8. **Strategic Funding Points**
   Dedicated external funds
9. **Demonstration Projects**
   Experimentation
   Technology Development
   Policy and Institutional learning

Brown and Clarke (2007) found that issue champions worked in a loose network of individual representatives from across government, academia, community and the land development sectors collectively pursuing change over a sustained period of time. The other enabling context variables relate to the level of socio-political capital for protecting waterway health, opportunity for strategic external funding avenues, and the establishment of bridging organisations to bring scientists and industry practitioners together. This interplay of associations and networks helped to formalise the objectives of improving stormwater quality, increasing large developers' receptivity to new practices in the marketplace, and facilitate the development of strategic capacity building tools that includes methods for envisioning future water management scenarios, water quality modelling software, and innovative design guidelines. This framework of enabling context variables was used by Tan and Wong (2009) in developing a strategy to institutionalise the practices of the Waterways City in Singapore.

Community acceptance and broad political support for WSUD is fundamental if it is to be implemented faster and if industry's technical capacity and ingenuity in complex urban environments is to be improved. However a major political shift cannot be simply assumed as providing the key transitioning ingredient, at least in the short to medium term. Keath and Brown's (2009) study of Brisbane, Australia revealed that, while the city was moving towards the Waterways City with many of the transition qualities as listed above, the political response to the crisis of extreme water scarcity represented a significant reversion to the earlier Water Supply management regimes rather than moving towards a Water Cycle City. Thus despite strong niche activities to institutionalise water cycle practices, the regime was unable to take advantage of the window of opportunity that the very large water scarcity driver opened for progressing further toward the Water Cycle City.

In fact, the conclusion of the study suggested that the lack of preparation for these extreme events may result in a backward step as tried and trusted solutions are favoured in times requiring quick, fail safe responses. Some of these preparation activities include, among others, active niche building, developing transitions strategies for multiple future scenarios, producing scientifically robust alternative solutions, building strategic links into management regimes, fostering political receptivity through high-level stakeholder engagement, and entrepreneurial leadership.

While no research currently exists on the experiences of developing nations in their urban water management evolution, it would seem important to focus on facilitating multiple opportunities for them to avoid the lock-in developed nations are now facing. A first step would be comprehensive consideration of system design and infrastructure selection before it 'goes in the ground.' But as the transitions framework implies, unless the socio-political drivers of amenity and environmental protection, recognition of natural resource limits and intergenerational equity are present in the hydro-social contract, it is unlikely that the urban water management system will respond with a service delivery model meeting these needs. With secure water supply access and public health the pressing need in developing countries, it may be unreasonable to expect that these additional goals and the more complex arrangements needed to implement them should be expected when basic human needs are not met. Yet this may also be the opportune context for facilitating the widespread implementation and use of decentralised practices. More insight is needed to explore the hydro-social contract and resulting opportunities for Water Sensitive Cities in developing nations.

3.6 CONCLUSION

These gaps in knowledge are only relatively recently beginning to be addressed, with researchers from a variety of disciplinary backgrounds attempting to define the research agenda for examining the socio-technical attributes and institutional arrangements needed to underpin the paradigm shift highlighted (Huitema *et al.*, 2009; Blomquist *et al.*, 2004; Geels 2004), providing the conceptual development of the theory to examine the urban water sector (Grin *et al.*, 2010; Pahl-Wostl *et al.*, 2009), empirical studies of past change processes (Brown and Clarke 2007; van der Brugge *et al.*, 2005), explorations of factors underpinning these changes (Huitema and Meijerink 2009; Olsson *et al.*, 2006) and frameworks and tools to aid in the identification of strategies to progress the paradigm shift (Loorbach 2010; Wong and Brown 2009). Research on Australian cities experiences highlight that active niche building, preparing transitions strategies, developing multiple future scenarios, scientifically robust alternative practices, and high level stakeholder engagement are key to progressing to the Waterways, Water Cycle and Water Sensitive Cities. The urban water scholarship is only just beginning to address the paucity of insight into the characteristics of enabling socio-technical forces that support the emergence, co-existence and subsequent over-riding of old institutional and technological path dependent practices. The challenge for scholars will be to provide this insight and knowledge in relevant and easily digestible forms for everyday practice and working contexts of the urban water professionals responsible for implementing a paradigm shift. Some of this challenge is being addressed through the research of the Monash University, Centre for Water Sensitive Cities (see www.watersensitivecities.org.au), and other programs such as SWITCH-Managing Water for the City of the Future (http://www.switchurbanwater.eu/).

REFERENCES

Arthur, W.B. (1989) Competing Technologies, Increasing Returns, and Lock-in by Historical Events. *Econ. J.* **99**, 116–131.

Ashley, R.M. (2005) Sustainable Disposal of Domestic Sanitary Waste. *J. Envion. Eng.* **131**(2), 206–215.

Barnett, J. and O'Neill, S. (2010) Maladaptation. *Global Environ. Chang.* **20**(2), 211–213.

Backhurst, M., Berke, P., Crawford, J., Day, M., Dixon, J., Ericksen, N. and Laurian, L. (2004) Evaluating plan implementation: a conformance-based methodology. *J. Am. Plann. Assoc.* **70**(4), 471–481.

Blomquist, W., Heikkila, T. and Schlager, E. (2004) Building the agenda for institutional research in water resource management. *J. Am. Water. Resour. As.* **40**, 925–936.

Brandes, O.M. and Kriwoken, L. (2006) Changing perspectives-changing paradigms: taking the "soft path" to water sustainability in the Okanagan Basin. *Can. Water Res. J.* **31**(2), 75–90.

Brown, R., Beringer, J., Deletic, A., Farrelly, Fletcher, T, M., Grace, M., Hart, B.T., Hatt, B., Keath, N., Lake, S., Lynch, A., Mac Nally, R., Mitchell, G., Mfodwo, K., Morison, P., Stubbs, W., Tapper, N., van de Meene, S. and Wong, T. (2007) Monash University Submission to the Review of the Metropolitan Water Sector: 'Moving to the Water Sensitive City: Principles for Reform', *Public Submission to the Victorian*

Competition & Efficiency Commission review of the metropolitan Water Sector, 5th October 2007.

Brown, R.R. and Clarke, J.M. (2007) Transition to Water Sensitive Urban Design: The Story of Melbourne, Australia. Facility for Advancing Water Biofiltration, Monash University, Melbourne.

Brown, R.R. (2008) Local Institutional Development and Organisational Change for Advancing Sustainable Urban Water Futures. *Environ. Manage.* **41**(2), 221–233.

Brown, R.R. and Farrelly, M.A. (2009) Delivering sustainable urban water management: a review of the hurdles we face. *Wat. Sci. Technol.* **59**(5), 839–846.

Brown, R., Keath, N. and Wong, T. (2009) Urban Water Management in Cities: Historical, Current and Future Regimes, *Wat. Sci. Technol.* **59**(5), 847–855.

Dawson, R. (2007) Re-engineering cities: a framework for Adaptation to Global Change. *Philos. T. R. Soc. A.* **365**, 3085–3098.

Duffy, A. (2009) Summary Report: Benchmarking Assessment of SWITCH Program Partner Cities. Reported to *The Global City Water Futures Summit*, 7th to 10th October 2009, Delft, The Netherlands, on behalf of UNESCO-IHE SWITCH–Managing Water for the City of the Future Program, unpublished.

Folke, C. (2006) Resilience: The emergence of a perspective for socio-ecological systems analysis. *Global Environ. Chang.* **16**(3), 253–267.

Geels, F.W. (2004) From sectoral systems of innovation to socio-technical systems - Insights about dynamics and change from sociology and institutional theory. *Res. Policy.* **33**, 897–920.

Gleik, P.H. (2003) Water Use. *Annu. Rev. Env. Resour.* **28**, 275–314.

Grin, J., Rotmans, J., Schot, J., Geels, F. and Loorbach, D. (Eds.) Transitions to Sustainable Development: New Directions in the Study of Long Term Transformative Change. New York, Routledge.

Harding, R. (2006) Ecologically sustainable development: origins, implementation and challenges. *Desalination.* **187**, 229–239.

Huitema, D. and Meijerink, S. V. (2009) Water policy entrepreneurs: a research companion to water transitions around the globe. Edward Elgar, Cheltenham.

Huitema, D., Mostert, E., Egas, W., Moellenkamp, S., Pahl-Wostl, C. and Yalcin, R. (2009) Adaptive Water Governance: Assessing the Institutional Prescriptions of Adaptive (Co-) Management from a Governance Perspective and Defining a Research Agenda. *Ecol. Soc.*, **14**, 19.

Keath, N. and Brown, R. (2009) Extreme Events: Being Prepared for the Pitfalls with Progressing Sustainable Urban Water Management, *Wat. Sci. Technol.* **59**(7), 1271–1280.

Loorbach, D. (2010) Transition Management for Sustainable Development: A Prescriptive, Complexity-Based Governance Framework. *Governance.* **23**, 161–183.

Lundqvist, J., Turton, A. and Narain, S. (2001) Social, institutional and regulatory issues. In *Frontiers in urban water management: Deadlock or hope.* (eds. C. Maksimovic and J.A. Tejada-Guilbert) IWA Publishing, Cornwall.

Maksimovic, C. and Tejada-Guibert, J.A. (Eds.) (2001) Frontiers in Urban Water Management: Deadlock or Hope. IWA, Cornwall.

Marsalek, J., Rochfort, Q. and Savic, D. (2001) Urban water as a part of integrated catchment management. In *Frontiers in urban water management: Deadlock or hope.* (eds. C. Maksimovic and J.A. Tejada-Guilbert) IWA Publishing, Cornwall.

Meadowcroft, J. (2005) Environmental political economy, technological transitions and the state. *New Polit. Econ.* **10**(4), 479–498.

Milly, P.C.D., Betancourt, J., Falkenmark, M., Hirsch, R.M., Kundzewicz, Z.W., Lettenmaier, D.P. and Stouffer, R.J. (2008) Climate change - Stationarity is dead: Whither water management? *Science*. **319**, 573–574.

Mitchell, B. (2005) Integrated water resource management, institutional arrangements, and land-use planning. *Environ Plann. A*, **37**, 1335–1352.

Mitchell, V.G. (2006) Applying integrated urban water management concepts: A review of Australian experience. *Environ. Manage.* **37**, 589–605.

Newman, P. (2001) Sustainable urban water systems in rich and poor cities – steps towards a new approach. *Wat. Sci. Technol.* **43**, 93–99.

Olsson, P., Gunderson, L.H., Carpenter, S.R., Ryan, P., Lebel, L., Folke, C. and Holling, C.S. (2006) Shooting the rapids: Navigating transitions to adaptive governance of social-ecological systems. *Ecol. Soc.* **11**, 21.

Pahl-Wostl, C., Sendzimir, J. and Jeffrey, P. (2009) Resources Management in Transition. *Ecol. Soc.* **14**(1), 46.

Saleth, R.M. and Dinar, A. (2005) Water institutional reforms: theory and practice. *Water Policy*, **7,** 1–19.

Smit, B. and Wanadel, J. (2006) Adaptation, adaptive capacity and vulnerability. *Global Environ. Chang.* **16**(3), 282–292.

Tan, N.S. and Wong, T.H.F. (2009), Active, Beautiful and Clean (ABC) Waters Programme: towards sustainable stormwater management in Singapore, keynote, proceedings of the 6th International Water Sensitive Urban Design Conference and Hydropolis #3, 5–8 May, 2009.

The Barton Group. (2005) Australian Water Industry Roadmap: A Strategic Blueprint for Sustainable Water Industry Development. Report of The Barton Group, Coalition of Australian Environment Industry Leaders, May 2005. Available at www.bartongroup.org.au

van der Brugge, R., Rotmans, J. and Loorbach, D. (2005) The transition in Dutch water management. *Reg. Environ. Change*, **5**, 164–176.

Walker, W. (2000) Entrapment in large technology systems: institutional commitment and power relations. *Res. Policy*. **29**, 833–846.

Wong, T.H.F. (2006) Chapter 1 – Introduction. In *Australian Runoff Quality: A guide to Water Sensitive Urban Design*. (ed T.H.F. Wong) Engineers Australia, Canberra.

Wong, T.H.F. and Brown, R.R. (2009) The water sensitive city: principles for practice. *Wat. Sci. Technol.* **60**, 673–682.

Wong, T.H.F. and Brown, R.R. (2011). Water Sensitive Urban Design, *In* Grafton, Q. and Hussey, K. (Eds.), '*Water Resources, Planning and Management: Challenges and Solutions*', Cambridge University Press.

Chapter 4

Effectively managing the transition towards restorative futures in the sewage industry: a phosphorus case study

Cynthia Mitchell, D. Fam and D. Cordell
University of Technology Sydney, Institute for Sustainable Futures, Sydney, Australia

4.1 INTRODUCTION

The water and sewage industry is at a transformation point. Climate change impacts, changing hydrological conditions, population growth, resource scarcity, aging infrastructure, economic constraints in financing large scale systems, and changing expectations for water quality (e.g. European Union water directives) all challenge existing planning parameters for developing and extending conventional centralized water and sewage systems.

While the resource intense nature of centralized water-based sanitation systems will struggle to adapt to these future uncertainties, there is a growing recognition within the water sector that an environmentally, economically and socially sustainable sanitation system requires sewage to be viewed as a set of resources to be recovered, recycled and reused (water, energy, nutrients) rather than a waste product to be treated to successively higher standards before release to the environment. Whilst such concepts have been core to the eco-sanitation advocates for many years, what is different now is that the concepts are gaining traction in the mainstream, large-scale end of the industry. For example, the 2009 International Water Association Leading Edge Technology conference, held as part of Singapore International Water Week, attracting up to 10,000 delegates from across the globe, opened with a workshop explicitly focused on carbon and nutrient recovery (see www.siww.com.sg). The impact of these realizations on the form, scale and operation of the sewage sector could be enormous (see Figure 4.1).

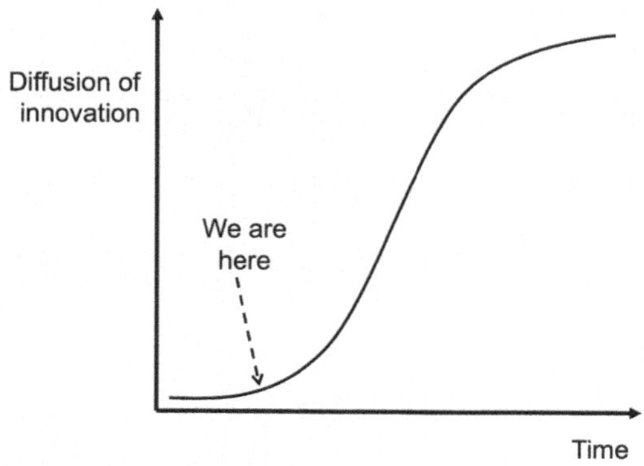

Figure 4.1 Graphic representation of the scale of change in the sewage sector in coming decades (After Rodgers, E.M. 2003 Diffusion of Innovations, 5th edn, Free Press, New York)

To date, resource recovery from sewage has largely focused on water due to increasing concerns of water scarcity, exacerbated by climate change impacts, urbanization and population growth. The national water industry association in Australia recognizes this and the implications for infrastructure in its vision for a sustainable urban water future: 'Given the need to maximize the efficient use of recycled water, it is highly likely that the days of extending sewage collection systems over ever-increasing distances to be connected to coastal sewage treatment plants are coming to an end.' (Water Services Association of Australia, 2009, p. 7). Elsewhere, nutrient recovery has emerged in areas (such as The Netherlands, Germany, Sweden) where changes in sewage management practices are being driven by concerns about contributions to nutrient imbalances, eutrophication, discharge of pharmaceuticals and a loss of biodiversity in receiving water bodies.

By shifting the paradigm from 'removal' to 'recovery and reuse' there is the potential to move beyond 'sustainability' toward developing 'restorative' systems (McGee et al., 2008) of water and sewage management. Restorative systems aim to have positive economic, social and ecological impacts, and are necessary for remediating historical effects of conventional sewage systems and in making up for areas that will not reach the baseline goal of sustainability (Mitchell 2008).

Innovative thinking across various dimensions is required to move toward restorative systems and is occurring globally with emergence of four significant themes:

4.1.1 New costing perspectives

There is an increasing interest in costing approaches that expand beyond a narrow focus on least financial cost to a single stakeholder, and encompass broader costs and benefits and broader stakeholders. One such approach is integrated resource planning (Swisher 1997, Vickers 2001), which seeks the lowest cost to society across the life of the infrastructure whilst providing socially and environmentally preferable outcomes (Mitchell et al., 2007). Since water and sewage management are generally provided by public funds, least cost investment in water and sewage infrastructure and its operation, frees up funds for investing in other public infrastructure, like schools and hospitals. Another emerging approach is 'value based' evaluation to design and implement systems that have the most value. This shift in perspective in valuation processes is occurring globally (e.g. The Vancouver Valuation Accord). Another is the increasing interest in bringing externalities inside the decision-making process, through sustainability accounting (see Yarra Valley Water, 2009, p. 12), assigning value to externalities or deliberative engagement (see below). All of these economic evaluation approaches have the potential to radically change what decisions are made by water authorities in developing water and sewage management systems.

4.1.2 Participatory, deliberative decision making in developing water systems

As the adoption of distributed systems and nutrient recovery presents a paradigm shift in wastewater management, community engagement is essential for the introduction of alternative socio-technical systems. By adopting 'representative and deliberative' processes of engagement in decision making, the aim is not only to select participants who are representative of the society of concern but also provide outcomes from the process that will impact the decisions of authorities. Not only does this approach provide the potential for delivering 'well-informed, fair and equitable decisions' (Fung & Wright 2003) in the water sector but also of providing much needed public support for new approaches to water and sewage management and the associated institutional arrangements needed to support their introduction. These approaches are closely linked to those that expand notions of costing.

4.1.3 Shifting from a resource focus to service focus

While conventional business thinking takes an output focus and is concerned with supplying a commodity, a shift in the sustainable business arena is occurring toward being outcome focus and supplying a service rather than supplying a resource (Dunphy, Griffiths & Benn 2003). This conceptual shift in thinking is providing significant gains in economic, environmental and social outcomes while improving financial performance (See Mitchell et al., 2007 for further examples).

4.1.4 Systemic thinking & synergies between energy, water reuse and nutrient recovery

Systemic thinking is increasingly being acknowledged as critical for sustainable management of our natural resources on a local, regional or global scale (see for example, the outputs of water-oriented multi-party international research programs focused on social learning (SLIM [online]), earth systems science (ESSP [online]), and sustainability (Urban Water [online]). However the institutional and physical structures created to manage natural resources over the last decades do not reflect a translation of systems thinking into practice. A good example of this is the nature of most centralized water and sanitation systems and the potential connection to the global food production and consumption system. While both the sanitation and agricultural systems have the potential to take advantage of synergies between material flows, such as reuse of wastewater fractions in food production, rarely does either system integrate and make use of the potential resources available. Instead, in an energy and resource intense way, the sanitation and agricultural sectors are continually sourcing new resources of water and nutrients (Cordell 2008).

A restorative future for the sanitation and water sector would encompass all these shifts, and provide a starting point for backcasting (Dreborg 1996, Mitchell and White 2003) to the present to identify alternative paths for infrastructure planning and investment, in developed and developing countries alike.

There is little doubt that the embedded nature of the sanitation system, with intertwined components of physical assets, organizations, institutions and social habits of practice, creates significant barriers to system innovation (Fam et al., 2009) and integrating material flows across the sanitation and agricultural sector. But the critical nature of resource depletion (nutrients) and scarcity (water, energy, nutrients) means there is greater acknowledgement by the water sector (e.g. the Victorian Water Industry Association's annual sustainability seminar in 2009 had the title: 'Responding to a water, carbon and nutrient constrained future') of the need for sustainable resource recovery from sewage. Today the question is therefore not *whether* technological change should occur but *how* to strategically manage a transition toward sustainable means of resource recovery. The water sector has the opportunity and is in the position to facilitate such a shift toward technological change and sustainable reuse of constituents in sewage.

4.2 MANAGING A TRANSITION TOWARD RESTORATIVE FUTURES

In considering the potential for system change, the weight of the past creates significant barriers to transformation of the water sector and transitioning toward a restorative future. System innovation or transition is generally defined as fundamental transformations in the way societal functions, such as transportation, communication, housing, waste management, are fulfilled (Elzen and Wieczorek 2005). The emphasis is on the co-evolution of technical and societal change as opposed to incremental change which is characterized primarily by changes in technology. Many of the barriers to radical change and system innovation in sewage management relate to the embedded nature of the centralized sanitation system which has co-evolved over the last century to become a highly interdependent, complex system of institutions, organizations, material infrastructures, technologies and social habits of practice (Hughes 1987) that make it challenging to introduce radical alternatives to the mainstream even if those alternatives may provide more sustainable outcomes.

As with many other large-scale infrastructural systems, such as energy supply and transportation systems, the sanitation system is weighed down by enormous sunk infrastructural costs, embedded institutional and regulatory structures and a life span of physical assets that can last 100 years or more. Shifting the trajectory of sanitation systems toward sustainability is therefore complicated – the investments and decisions made in the past create a tendency to incrementally add to and extend the existing system rather than invest in innovations with radically different sustainability potentials. Although understanding of sustainability is constantly evolving, wastewater design and management processes are aligned to 20th century engineering traditions (Guest et al., 2009) and characterized by incremental change along existing paths and trajectories rather than radical innovation.

While many authors have discussed the difficulties of changing socio-technical systems (Dosi 1982; Metcalfe 1997; Rip & Kemp 1998; Rotmans, Kemp & Asselt 2001) there is increasing interest from policy makers, NGO's and industrial firms in system innovation and the sustainability benefits the transition may offer. This growing interest in system innovation is reflected in a shift in research of technology and sustainable development, from the level of single technologies to socio-technological regimes (Kemp, Schot & Hoogma 1998; Rip & Kemp 1998; Smith 2003). In the 1980s, effort was focused on solving environmental problems with cleaner product and process innovations that were developed in conjunction with end-of-pipe solutions (Geels 2006; Smith 2003). Although this incremental step to technological change has, in cases, led to significant improvements in environmental efficiency, substantial leaps in efficiency required to reach the demanding goals of restorative systems and long term sustainability in the water sector are only possible through system innovation or the large scale transformation of how societal functions such as sewage management are fulfilled (Elzen & Wieczorek 2005).

Introducing system innovation and cost effective resource recovery requires understanding socio-technical change, how it has occurred in the past and how we may potentially manage a transition toward sustainability in the future. Historically, socio-technical systems have involved multiple actors, factors and levels in system change. For example, the transformation of the sanitation system in many industrialized countries from the use of cesspools to sewer systems was a co-evolutionary process. The transition was not caused by a single factor or the introduction of a technological breakthrough but rather interplay of factors that influenced each other at varying levels. The co-evolution of technology and society in the development of centralised sanitation was a relationship between multiple factors (technical, regulatory, societal and behavioural), actors (users/consumers, government & private enterprise) and levels (macro, meso and micro) (Geels 2005).

Transition theorists argue that technological transitions and radical shifts in established socio-technical regimes such as wastewater management occur with interactions and alignment between processes at different levels – a multi-level perspective (Geels 2002; Geels & Schot 2007; Rip & Kemp 1998). For example, socio-technical landscape factors such as water scarcity coupled with urban growth have the potential to destabilize established socio-technical regimes (configurations of technologies, institutional structures, rules and norms) providing opportunity for new socio-technical configurations to occur at the level of the technological niche (radical technologies and practices). At the same time the path dependent characteristics of established socio-technical regimes may limit the emergence of radically different technologies and user practices (Geels 2002; Rip & Kemp 1998; Walker 2000).

Broad landscape drivers such as water scarcity and nutrient and energy constraints have triggered a number of water authorities and research institutes (e.g. the Melbourne water authorities (such as Yarra Valley Water), German Technical Corporation (GTZ), UNESCO-IHE, Dutch Foundation of Applied Water Research (STOWA)) to seriously consider adopting resource recovery systems such as urine diversion, a radical innovation in sewage management.

Although urine diversion is seen as radical today, it has been practiced since ancient times in cultures spanning from Greece to Asia. The modern water-based version of the urine diversion (UD) system was invented and emerged during the early 1990s, where waste streams are separated at the source at the household level with a urine diversion toilet. Due to the nutrient rich nature of urine, which consists of phosphorus, nitrogen and potassium, it is collected at the household level, stored, sanitised and reused on-site as a substitute for chemical fertilisers or centrally collected by the municipality in conjunction with local farmers before being transported for storage, sanitization and reuse in agricultural applications (see Johansson, M., H. Jonsson et al., 2000) for further details of Swedish experiences).

To practically implement resource recovery systems at any significant scale and deliberately manage a transition toward a sustainable or restorative future in the water sector, will undoubtedly involve a co-evolution of technological and mutually reinforcing institutional and socio-cultural transformation. For example, for the introduction and development of small scale systems of nutrient recovery such as urine diversion systems (UD):

- new regulations need to be developed (e.g. akin to the World Health Organisation guidelines for wastewater reuse (WHO 2006),
- risks and responses considered (e.g. in the development of decentralised or distributed systems),
- institutional arrangements managed (e.g. service teams constructed),
- innovative decision-making processes implemented (e.g. deliberative, participatory processes),
- interpretations of personal responsibility articulated (e.g. behavioural change, socio-cultural habits and practices),
- pricing and payment structures constructed (e.g. feed-in tariffs, tax incentives) and
- markets stimulated (e.g. for nutrients and new technologies).

In practical terms, experimentation means piloting, demonstrating, monitoring, evaluating and providing opportunity for learning across these multiple dimensions and within different application domains.

Technological niche experiments act as domains temporarily protected from harsh selection environments through mechanisms such as tax exemptions, investment grants and other forms of 'protection'. The niche environment is therefore fundamental in transitioning toward a more sustainable state of resource recovery and can be used as a strategic location for learning, building new social networks, improving the innovation and broader diffusion of the concept.

4.3 ADOPTING STRATEGIC NICHE MANAGEMENT TO IMPLEMENT RADICAL INNOVATION IN RESOURCE RECOVERY AND REUSE

The introduction of sustainable technologies is challenging in the face of existing systems characterized by lock-in and resistance to change (Unruh 2000), such as those represented by the global water and sanitation sector. Strategic Niche Management (SNM) therefore emerged as an area of research through the observation that many sustainable innovations fail to 'take off' in spite of their significant environmental potential. The framework of SNM is a rapidly growing area of research revolving around experimental-based learning of the technological, social and environmental possibilities and constraints of innovation (Caniels & Romijin 2008). 'Niche' is defined here as a protective space or 'incubation' area where actors can experiment with an innovation and learn about the potential barriers and benefits of system change (Raven 2005; Rip & Kemp 1998).

The SNM framework has been applied to various case studies in the area of sustainable innovation such as biogas plants and biomass firing (Raven 2005), sustainable transport (Hoogma et al., 2002) and organic food (Smith 2006) to name a few. The analysis of case studies tends to be of experiments, demonstration projects and pilot plants with scholars exploring the challenges, opportunities and the lessons learned. As Smith suggests this is a very different approach to researching transitions toward sustainability, the conventional approach being to study unsustainability and recommend reforms for mitigating these problems. The more novel approach of SNM explores '... innovative experiments in alternative, sustainable technological niches, drawing lessons from the challenges they face in the context of a dominant unsustainable technological regime.' (Smith 2003, p. 128).

Until now SNM has been a tool for analyzing past experiences: constructive applications of SNM are rare and limited. Raven (Raven 2005) notes that SNM has primarily been used for improving the design of experiments, evaluating policies in the past, using SNM as a scenario development tool and to design future policies on niche management. In practical terms, the SNM framework offers insight into how radical niche based technologies such as resource recovery and reuse systems might be designed and introduced to contribute to shifting practices away from resource intense infrastructures such as the dominant regime of centralized sanitation. Although we do not presume to present a comprehensive account of transition theory or the intricacies involved in operationalizing alternative technological experiments using the SNM framework, in practice decentralized and distributive systems will undoubtedly play a key role in transitioning toward a restorative futures of resource recovery as they provide a viable means for experimentation and *learning by doing* at a relatively low risk and low cost (Mitchell 2008).

4.4 PHOSPHORUS SECURITY: A CASE FOR TRANSITION

4.4.1 A new global challenge: from phosphorus pollution to phosphorus scarcity

Whilst it is widely acknowledged that addressing future availability of energy and water resources will be critical for meeting future demands of a growing global population, the need to address the issue of limited phosphorus availability is at least as serious, and not yet widely recognized. Awareness and management of environmental challenges related to phosphorus have typically been associated with water pollution and eutrophication. Today we are on the brink of a new global understanding: global phosphorus scarcity and the serious threat it poses to future food production (Cordell et al., 2009a).

Phosphorus (P), together with Nitrogen (N) and Potassium (K), is a critical element for plant and animal growth, and therefore food production. There is no substitute for phosphorus in food production and it cannot be manufactured, hence its significance to humanity (Cordell et al., 2009a).

Historically, crop production relied on natural levels of soil phosphorus with the addition of organic matter like manure and in parts of Asia, human excreta ('nightsoil') (Mårald 1998). To keep up with rapid population growth and food demand in the 20th century, concentrated mineral sources of phosphorus were discovered in guano and phosphate rock and applied extensively (Brink 1977; Smil 2000). Chemical fertilizers (containing N, P, K) contributed to feeding billions by boosting crop yields between 1950–2000 (IFPRI 2002). Additions of phosphorus fertilizer are essential for maintaining high crop yields and replenishing soil with what is taken away by harvested crops, especially in agricultural systems with naturally phosphorus deficient soils. Modern agriculture today is dependent on fertilizers derived from phosphate rock: around 170 million tonnes of phosphate rock are mined and traded each year (containing around 23 million tonnes elemental P), 90% of which is used for the production of phosphate fertilizers (and to a lesser extent animal feed and food additives).

Yet phosphate rock, like oil, is a non-renewable resource and approximately 50–100 years remain of current known reserves. Further, a peak in global production – peak phosphorus – is estimated to occur by 2035 (Cordell et al., 2009a). After the peak, supply will decrease year upon year, constrained by economic and energy costs, despite rising demand. While the exact timing of the peak may be disputed, there is general consensus among industry and scientists that the quality of remaining reserves is declining and that cheap fertilizers will become a thing of the past in the long term (IFA 2006; Smil 2002; Stewart Hammond & Kauwenbergh 2005). Increasing energy and other resources are required to mine, process and extract the same nutrient value from phosphate rock.

While all farmers need access to phosphorus fertilizers, just five countries control around 90% of the world's remaining reserves, including China, the US and Morocco (Jasinski 2009). The period 2007–2008 saw an unprecedented 800% spike in the price of phosphate rock (Minemakers Limited 2008; World Bank 2009) that was unforeseen by most of the world's farmers and policy-makers. China has the largest reported reserves, yet in 2008 it imposed a 135% export tariff on phosphate, effectively banning any exports in order to secure domestic supply for food production (Fertilizer Week 2008). The US is running out of its domestic high-grade reserves and increasingly importing rock from Morocco to process into high analysis fertilizer for sale on the world market. This is geopolitically sensitive as Morocco currently occupies Western Sahara and controls its vast phosphate rock reserves. Trading with Moroccan authorities for Western Sahara's phosphate rock is condemned by the UN, and importing phosphate rock via Morocco has been boycotted by several Scandinavian firms (Corell 2002; WSRW 2007). Further, in a carbon-constrained future, shipping millions of tonnes of phosphate rock and fertilizers around the globe may no longer be either appropriate or feasible.

The demand for phosphorus is expected to increase over the long-term due to increasing world population, preferences towards more meat- and dairy-based diets (which demand more phosphorus) in emerging economies, and increasing demand for non-food crops such as biofuel crops. Further, current market demand for phosphorus fertilizers only represents those farmers with sufficient purchasing power. An additional 'silent' demand is also present from poor farmers with phosphorus-deficient soils, particularly in Sub-Saharan Africa (Cordell et al., 2009b).

Despite this growing demand coupled with increasing phosphorus scarcity, there are no alternatives to phosphate rock on the market today that could replace it on any significant scale. While small scale operations and trials of phosphorus recovery from excreta and other waste streams exist, commercialization and implementation on a global scale could take decades to develop and significant adjustments to institutional arrangements will be required to support these infrastructural changes (Cordell et al., 2009a).

However the current use of phosphorus is extremely inefficient: approximately 80% of the phosphorus mined in phosphate rock for food production never actually reaches the food on our forks (Cordell et al., 2009a). Phosphorus is lost at all key stages of the food production and consumption chain: from mining to fertilizer application, crop uptake and harvest, food processing distribution and consumption by humans and animals. Close to 100% of phosphorus consumed in food is excreted (Jonsson et al., 2004). Human excreta (urine and faeces) are renewable and readily available sources of phosphorus. Urine is essentially sterile and contains plant available nutrients (P, N, K) in the correct ratio. Indeed, the global population generates around 3 million tones of elemental phosphorus in urine and faeces each year which typically ends up in waterways if not intentionally recovered (Cordell et al., 2009a). Given half the world's population lives in urban centres and urbanization is set to increase, cities are becoming phosphorus 'hotspots'. This means there is a substantial opportunity to consider localized phosphorus recovery. Urine alone represents the largest single source of phosphorus emerging from cities. Recirculating urban nutrients such as urine back to agriculture therefore presents an enormous opportunity for the future (see Figure 4.2). Further, sourcing phosphorus from more local, renewable sources

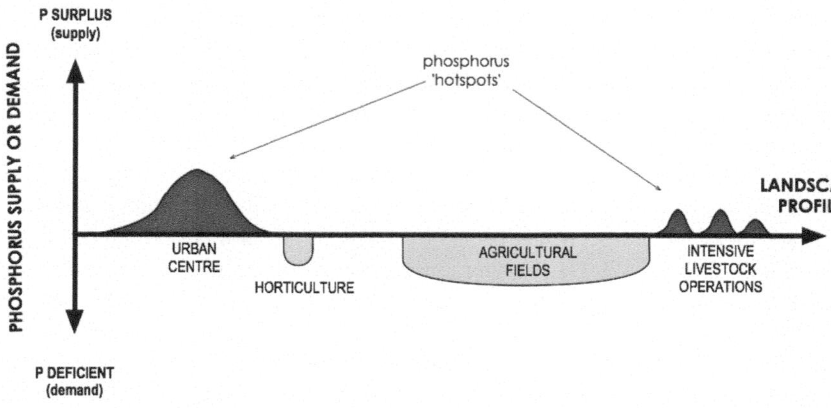

Figure 4.2 Spatial profile of an urban-rural landscape – indicating that while agricultural and horticultural fields demand continual phosphorus fertilizers, cities are 'phosphorus hotspots' of food waste and human excreta that could be productively utilized to meet some of the fertilizer demand. Source: (Cordell 2010)

(such as excreta or food waste) rather than depending on access to an increasingly scarce global commodity can increase communities' phosphorus security and potentially reduce wastage and energy consumption in the food system.

4.5 THE ROLE OF NICHE BASED APPROACHES TO RESOURCE RECOVERY IN A SUSTAINABLE PHOSPHOROUS FUTURE

While incremental system change, such as increasing the efficient use of phosphorus in agriculture is likely to be an important measure, optimizing current practices will alone not be sufficient to secure a sustainable phosphorus future. More fundamental changes will be required that seek to incorporate a range of supply- and demand-side measures, as depicted in Figure 4.3 (Cordell et al., 2009b).

For example, business-as-usual global demand for phosphorus can be substantially reduced through measures that seek to change diets to less meat and dairy (which require more fertilizer per unit of food nutrition), reduce losses in the food chain, and increase efficient use in agricultural and livestock sectors. On the supply side, the water sector is likely to play a critical role, together with the food and other sectors in recovering and reusing phosphorus from the entire food production and consumption system: from excreta to food waste to crop residues.

There are several socio-technical sanitation systems today that support the recovery and re-use of nutrients, from small-scale grey water systems for example, source-separating composting toilets, wetlands for recovering nutrients, to more large-scale high-tech recovery from mixed wastewater streams. If a key objective of nutrient recovery is to *reuse* the nutrients in food production (which is indeed argued in this chapter), then health, environmental and economic costs of the entire system from toilet to field become paramount. Reuse is often safer if sanitation service providers and urban planners avoid infrastructure that mixes human excreta with other wastewater streams such as industrial wastewater which often contain heavy metals and other toxic contaminants (for example through the use of decentralized and distributed sanitation systems). Further, if urine is separated from faecal matter in the toilet, the urine can be safely used through simple storage (WHO 2006). Urine has the potential to provide half the phosphorus required to fertilize cereal crops (Drangert 1998; WHO 2006). While reusing human excreta has been practiced in parts of the world for over 5000 years, Drangert (1998) suggests a 'urine-blindness' has prevented modern societies from tapping into this bountiful source of plant nutrients and thus there is still a need to legitimize the use of human excreta among water authorities in many parts of the world.

In addition to reuse of source-separated urine and faeces, examples of other nutrient recovery and reuse systems ranging from commercial operations to R&D

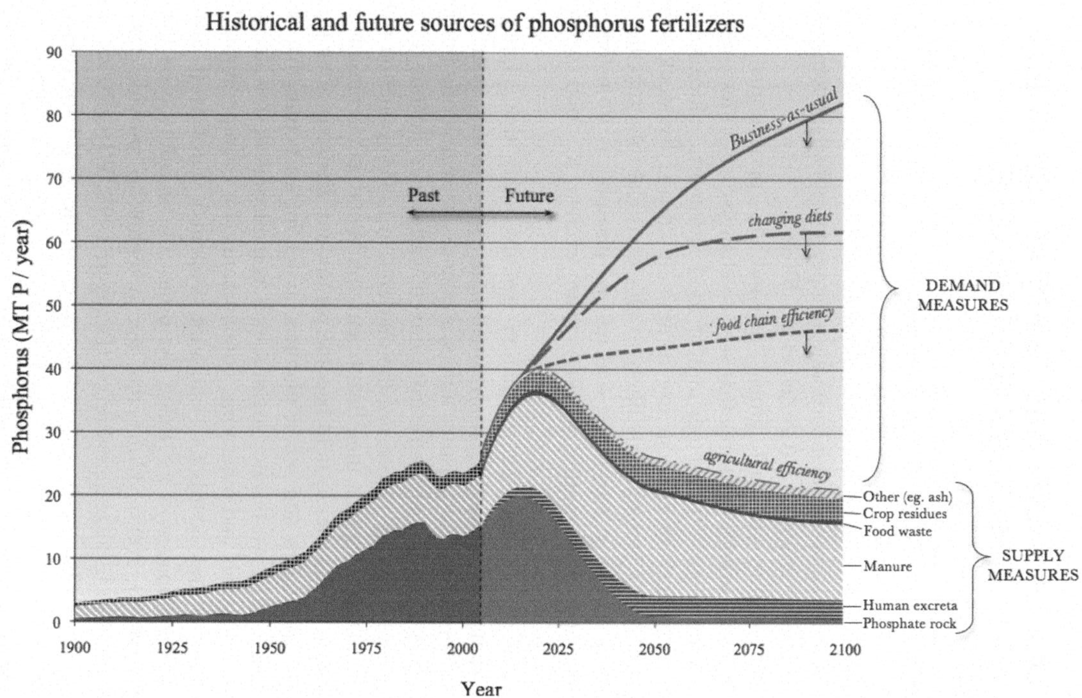

Figure 4.3 Closing the gap: meeting future global phosphorus demand for food security through a range of ambitious demand and supply-side measures (adapted from Cordell et al., 2009b)

phase include struvite recovery from wastewater treatment plants (Ostara 2009; Rahamana et al., 2009), use of ash from incinerated sludge (Schipper & Korving 2009), or a combination such as struvite precipitation from source-separated urine (Ganrot, Broberg & Byden 2009; Tilley et al., 2009). Indeed, Schenk et al. (2009) suggest there are over 30 processes for the recovery of phosphorus from wastewater. While such emerging initiatives are certainly on the increase around the world, they are far from the mainstream and are generally not operating within an overarching coordinated framework or strategy at a broader scale linked specifically to sustainable nutrient recovery, sanitation and food production (Cordell 2010). Hence there is still a need to investigate the most appropriate ways of recovering phosphorus in a given context (within a region, country, city) as it is likely that no one social-technical solution will meet all needs. Important criteria for consideration might include life-cycle energy costs, other resource inputs, spatial distribution between nutrient recovery system and end users, farmer's views and preferences regarding the product, effectiveness of the product as a fertilizer and so on.

Implementing (or even trialing in some instances) such demand and supply-side measures are severely hindered by a fragmented institutional setting. For example, there are no policies, regimes or institutions explicitly ensuring long-term accessibility and availability of phosphorus resources for global food security (Cordell 2010). Global phosphorus resources are by default governed by the market system, which alone is not sufficient to ensure all farmers have access to sufficient phosphorus resources for food production and ensure environmental protection in both the short and long-term (Cordell 2010).

There is a 'lack of fit' (Young 2002) or mismatch between the phosphorus cycle and the institutional arrangements governing phosphorus resources. There is a noticeable fragmentation between the different sectors that phosphorus flows through in the global food production and consumption system. For example, there is a mismatch between the agricultural sector where phosphorus is typically perceived as a fertilizer commodity and the water and sanitation sector where phosphorus is typically perceived as a pollutant in wastewater (Cordell 2010; Cordell et al., 2009a). Phosphorus scarcity is currently not a priority within any sector and has no institutional home. It is only when the phosphorus cycle is perceived as a whole system which includes connections between entities, that its importance becomes obvious (Cordell 2008). Similarly, and partly as a consequence of this institutional fragmentation between the sanitation and food sector, urine diversion and reuse systems have no institutional home (Cordell 2006).

There have already been significant R&D and piloting of UD systems to recover, recycle and reuse nutrients in urine to land applications. For example twenty years of experiential learning and applied research in Sweden have provided insights into mutli-dimensional challenges and opportunities of transitioning toward small scale nutrient recovery (Johansson et al., 2006). The variable success of UD systems in Sweden highlights the fact that radical innovation cannot be attempted by a simple technological fix but rather requires a radical change in the institutional, technological and social foundations of these systems (Fam et al., 2009; Cordell 2006), in other words a change in the wider socio-technical setting that structures the behaviour and decisions of a broad range of actors involved (Raven 2007).

In an unprecedented era of global environmental change, it is no longer sufficient or appropriate to take a single-sector approach to complex sustainability challenges. Achieving a sustainable phosphorus situation will require an integrated and collaborative approach to developing new policies, actors and partnerships to support the trialing and implementation of novel socio-technical systems. Proposed sustainability criteria for the goal of phosphorus security addresses social dimensions (such as farmer livelihoods), ecological dimensions (such as soil fertility), economic dimensions (such as cost-effectiveness of recovery systems), environmental dimensions (such as minimizing waste and pollution) (Cordell 2010).

Despite the presence of uncertainty and some degree of lack of consensus about the probable future of phosphorus, what is clear is that unless we intentionally change the way we source and use phosphorus, we will end up in a 'hard-landing' situation with increased phosphorus scarcity and phosphorus pollution, further fertilizer price fluctuations and increasing energy consumption. In order to achieve a preferred 'soft-landing' outcome, an integrated and globally coordinated approach to managing phosphorus is required. This is likely to require substantial change in both the physical and institutional infrastructure surrounding the sourcing and supply of phosphorus for food production. Such change cannot be achieved without the involvement of and innovation within the water and sanitation sector. SNM provides insights for the water sector and the process of system change. An SNM framework for supporting innovation and strategically designing experiments considers not only of the technological implications but also new institution and socio-cultural arrangements that will be necessary for successful diffusion of system change.

4.6 CONCLUSION

The water industry is at the start of a period of transformational change. In the coming decades, at least some of our water and sewage infrastructure needs to be restorative, operating in ways that aim for no net negative impact. The fundamental characteristics of sewage's key constituents (i.e. water, which is heavy; carbon, which is useful; and nutrients, particularly phosphorus, which is essential) mean that distributed systems, i.e. local scale infrastructure managed centrally, present particular opportunities for transitioning to restorative futures.

While there is no quick fix solution for current dependence on phosphorus fertilisers, there are a number of technologies and policy options that exist at various stages of development – from research to demonstration to implementation, and the water sector has the potential to play an important role in scaling up such concepts to recover phosphorus from urban wastewater streams.

Meeting these and other pressing challenges in the sector requires significant systemic innovation. Understanding and taking action for systemic innovation requires viewing the issue from multiple perspectives and taking a transdisciplinary approach in creating change. Adopting approaches that are inclusive of stakeholders (broad collaboration), are context dependent (situation/problem based), and support flexible and reflexive approaches to systems change (evolving methodology) (Wickson, Carew & Russell 2006). Transition management offers some useful insights on how to create such system-wide change.

ACKNOWLEDGEMENTS

This chapter was developed with assistance from the Water Environment Research Foundation, and first published as a contribution to 'Water Sustainability and International Innovation: The Baltimore Charter – A Transformation in Managing Water'. Edited by: Valerie I. Nelson, Jerry Stonebridge, Steve Moddemeyer. Copyright © Coalition for Alternative Wastewater. Treatment. September 2010.

REFERENCES

Brink, J. 1977, 'World resources of phosphorus', *Ciba Foundation Symposium*, vol. Sept 13–15, no. 57, pp. 23–48.

Caniels, M.C.J. & Romijin, H.A. 2008, 'Strategic niche management: towards a policy tool for sustainable development', *Technology Analysis and Strategic Management*, vol. 20, no. 2, pp. 245–266.

Cordell, D. 2010, *The Story of Phosphorus: Sustainability implications of global phosphorus scarcity for food security*, Doctoral thesis. Collaborative PhD between the Institute for Sustainable Futures, University of Technology, Sydney (UTS) and Department of Water and Environmental Studies, Linköping University, Sweden. Linköping University Press, ISBN 978-91-7393-440-4, Linköping, http://urn.kb.se/resolve?urn=urn:nbn:se:liu:diva-53430.

Cordell, D., Drangert, J.-O. & White, S. 2009a, *The story of phosphorus: Global food security and food for thought.* Journal of Global Environmental Change, 19(2009), p. 292–305.

Cordell, D., Neset, T.S.S., Drangert, J.-O. & White, S. 2009b, *Preferred future phosphorus scenarios: A framework for meeting long-term phosphorus needs for global food demand*, International Conference on Nutrient Recovery from Wastewater Streams Vancouver, 2009. Edited by Don Mavinic, Ken Ashley and Fred Koch. ISBN:9781843392323. Published by IWA Publishing, London, UK.

Cordell, D. 2008, *Phosphorus, food and 'messy' problems: A systemic inquiring into the management of a critical global resource*, in Proceedings of the 14th ANZSYS Australia New Zealand Systems Society Conference, ed David Cook, SECAU - Security Research Centre, Edith Cowan University, December 2008, Perth, pp. 1–15. ISBN 978-0-7298-0668-8.

Cordell, D. 2006, *Urine Diversion and Reuse in Australia: A homeless paradigm or sustainable solution for the future?*, February 2006, Masters Thesis, Masters of Water Resources & Livelihood Security, Department of Water & Environmental Studies, Linköping University, Linköping. Available: http://www.ep.liu.se/undergraduate/abstract.xsql?dbid=8310.

Corell, H. 2002, *Letter dated 29 January 2002 from the Under-Secretary-General for Legal Affairs, the Legal Counsel, addressed to the President of the Security Council*, United National Security Council, Under-Secretary-General for Legal Affairs The Legal Counsel.

Dreborg, K.H. 1996, 'Essence of backcasting,' *Futures* 28 (9): 813–828.

Dosi, G. 1982, 'Technological paradigms and technological trajectories: a suggested interpretation of the determinants and directions of technological change', *Research Policy*, vol. 11, pp. 147–162.

Dunphy, Griffiths & Benn 2003, *Organisational change for corporate sustainability, a guide for leaders and change agents in the future*, Routledge, London.

ESSP [online] Earth Systems Science Partnership http://www.essp.org/. Accessed 6 May 2011.

Elzen, B. & Wieczorek, A. 2005, 'Transition toward sustainability through system innovation', *Technological Forecasting and Social Change*, vol. 72.

Fam, D., Mitchell, C. & Abeysuriya, K. 2009, 'From removal to recovery: critical stakeholder involvement in implementing pilot projects of urine diversion systems in Sweden', paper presented to the *The first annual Asia-Pacific Science Technology and Society Conference - Our Land, Our Waters, Our People*, Griffith University, Brisbane, Australia.

Fam, D.M., Mitchell, C.A., Lopes, A. & Willetts, J.R. 2009, 'The challenge of system change: an historical analysis of Sydney's sewer systems', *Design Philosophy Papers*, vol. 3/2009, no. 1–14.

Fertilizer Week 2008, *Industry ponders the impact of China's trade policy*, in Thursday Markets Report, 24th April 2008, British Sulphur Consultants, CRU.

Fung, A. & Wright, E.O. 2003, 'Thinking about empowered participatory governance', in A. Fung & E.O. Wright (eds), *Deepening democracy: institutional innovations in empowered participatory governance*, Verso, London.

Ganrot, Z., Broberg, J. & Byden, S. 2009, *Energy efficient nutrient recovery from household wastewater using struvite precipitation and zeolite adsorption techniques: A pilot plant study in Sweden*, International Conference on Nutrient Recovery from Wastewater Streams. Edited by Ken Ashley, Don Mavinic and Fred Koch. ISBN: 9781843392323. Published by IWA Publishing, London, UK.

Geels, F. 2002, 'Technological transitions as evolutionary reconfiguration processes: a multi-level perspective and a case study', *Research Policy*, vol. 31, pp. 1257–1274.

Geels, F. 2005, 'Co-evolution of technology and society: The transition in water supply and personal hygiene in the Netherlands (1850–1930) - a casestudy in multi-level perspective', *Technology in Society*, vol. 27, pp. 363–397.

Geels, F. 2006, 'System innovation and transitions to sustainability: challenges for innovation theory', *SPRU 40th Anniversary Conference*, p. 14.

Geels, F. & Schot, J. 2007, 'Typology of socio-technical transition pathways', *Research Policy*, vol. 36, pp. 399–417.

Guest, J.S., Skerlos, S.J., Barnard, J.L., Beck, M.B., Daigger, G.T., Hilger, H., Jackson, S.J., Karvazy, K., Kelly, L., Macpherson, L., Mihelcic, J.R., Pramanik, A., Raskin, L., L Van Loosdrecht, Yeh, D. & Love, N.G. 2009, 'A new planning and design paradigm to achieve sustainable resource recovery', *Environmental Science and Technology*, vol. 43 no. 16, pp. 6126–6130.

Hoogma, R., Kemp, R., Schot, J. & Truffer, B. 2002, *Experimenting for sustainable transport: the approach of strategic niche management*, Spon Press, London and New York.

Hughes, T.P. 1987, 'The evolution of large technical systems', in W.K. Bijker, T.P. Hughes & T. Pinch (eds), *The social construction of technological systems: new directions in the sociology and history of technology*, MIT Press, Massachusetts.

IFA 2006, *Production and International Trade Statistics*, International Fertilizer Industry Association, Paris, available: http://www.fertilizer.org/ifa/statistics/pit_public/pit_public_statistics.asp (accessed 20/8/07).

IFPRI 2002, *Reaching Sustainable Food Security for All by 2020: Getting the Priorities and Responsibilities Right*, International Food Policy Research Institute, Washington.

Jasinski, S.M. 2009, *Phosphate Rock, Mineral Commodity Summaries*, US Geological Survey, January, 2009, available: http://minerals.usgs.gov/minerals/pubs/commodity/phosphate_rock/mcs-2009-phosp.pdf.

Jonsson, H., Stintzing, A.R., Vinneras, B., Eckerberg, K. & Salomon 2004, *Guidelines for the use of urine and faeces in crop production*, EcoSanRes, Stockholm Environment Institute, Stockholm.

Kemp, R., Schot, J. & Hoogma, R. 1998, 'Regime shifts to sustainability through processes of niche formation: the approach of strategic niche management', *Technology Analysis and Strategic Management*, vol. 10, no. 2, pp. 175–192.

Mårald, E. 1998, *I mötet mellan jordbruk och kemi: agrikulturkemins framväxt på Lantbruksakademiens experimentalfält 1850–1907*, Institutionen för idéhistoria, Univ, Umeå.

McGee, C.M., Mitchell, C.A. & Retamal, M.L. 2008, 'City limits: pushing boundaries in urban infill development', World Sustainable Building Conference, Melbourne, September 2008 in *Proceedings of the 2008 World Sustainable Building Conference*, (ed) Foliente, G., Luetzkendorf, T., Newton, P. & Paevere, P. www.sb08melbourne.com, Melbourne, pp. 889–897.

Metcalfe, J.S. 1997, 'On diffusion and processes of technological change', in G. Antonelli & N.D. Liso (eds), *Economics of structural and technological change*, Routledge, London, UK.

Minemakers Limited 2008, *ROCK PHOSPHATE PRICE ROCKETS TO US$200/TONNE*, ASX and Press Release, Perth.

Mitchell, C.A. 2008, 'Restorative water: beyond sustainable.' *Waste Management and Environment Magazine* July 2008, pp. 20–21.

Mitchell, C.A., Fane, S.A., Willetts, J.R., Plant, R.A. & Kazaglis, A. 2007, *Costing for Sustainable Outcomes in Urban Water Systems - A Guidebook*, Cooperative Research Centre for Water Quality and Treatment, Sydney.

Ostara 2009, *Ostara Nutrient Recovery Technologies*, Homepage, www.ostara.com.

Rahamana, M.S., Mavinica, D.S., Britonb, A.T., Zhangb, M., Fattaha, K.P. & Koch, F.A. 2009, *Numerical investigations of the hydrodynamics of the UBC MAP fluidized bed crystallizer* International Conference on Nutrient Recovery from Wastewater Streams. Edited by Ken Ashley, Don Mavinic and Fred Koch. ISBN: 9781843392323. Published by IWA Publishing, London, UK.

Raven, R. 2005, 'Strategic Niche Management for Biomass: a comparative study of the experimental introduction of bioenergy technologies in the Netherlands and Denmark', Eindenhoven University, Netherlands.

Rip, A. & Kemp, R. 1998, 'Technological change', in, *Human Choice and Climate Change*, vol. 2, Batelle Press, Colombus, OH, pp. 327–399.

Rodgers, E.M. 2003, Diffusion of Innovations, 5th edn, Free Press, New York.

Rotmans, J., Kemp, R. & Asselt, M.V. 2001, 'More evolution than revolution: transition management in public policy', *Foresight*, vol. 3, no. 1, pp. 15–31.

Schenk, K., Fahrni, H.-P. & Laube, A. 2009, *Phosphorous in Switzerland Material flows, targets, options, P-recovery*, World Resources Forum, 14th-16th September, 2009, Davos, Switzerland.

Schipper, W.J. & Korving, L. 2009, *Full-scale plant test using sewage sludge ash as raw material for phosphorus production*, International Conference on Nutrient Recovery from Wastewater Streams. Edited by Ken Ashley, Don Mavinic and Fred Koch. ISBN: 9781843392323. Published by IWA Publishing, London, UK.

SLIM [online] *Social learning for integrated water managing* http://sites.google.com/site/slimsociallearningforiwm/, accessed 6 May 2011.

Smil, V. 2000, 'Phosphorus in the Environment: Natural Flows and Human Interferences', *Annual Review of Energy and the Environment*, vol. 25, pp. 53–88.

Smil, V. 2002, 'Phosphorus: Global Transfers', in P.I. Douglas (ed.), *Encyclopedia of Global Environmental Change*, vol. 3 'Causes and consequences of global environmental change', John Wiley & Sons, Chichester, pp. 536–542.

Smith, A. 2003, 'Transforming technological regimes for sustainable development: a role for alternative technology niches', *Science and Public Policy*, vol. 30, no. 2, pp. 127–135.

Smith, A. 2006, 'Green niches in sustainable development: the case of organic food in the United Kingdom', *Environment and Planning C*, vol. 24, pp. 439–458.

Swisher, J.N., Jannuzzi, G.D.M. et al. 1997, Tools and methods for integrated resource planning: Improving energy efficiency and protecting the environment. Riso National Laboratory, Denmark, UNEP Collaborating Centre on Energy and Environment.

Stewart, W., Hammond, L. & Kauwenbergh, S.J.V. 2005, 'Phosphorus as a Natural Resource', in, *Phosphorus: Agriculture and the Environment, Agronomy Monograph No.46*, American Society of Agronomy, Crop Science Society of America, Soil Science Society of America, Madison, pp. 3–21.

Tilley, E., Gantenbein, B., Khadka, R., Zurbru, C. & Udert, K.M. 2009, *Social and economic feasibility of struvite recovery from Urine at the community level in Nepal*, International Conference on Nutrient Recovery from Wastewater Streams. Edited by Ken Ashley, Don Mavinic and Fred Koch. ISBN: 9781843392323. Published by IWA Publishing, London, UK.

Unruh, G.C. 2000, 'Understanding carbon lock-in', *Energy Policy*, vol. 28, pp. 817–830.

Urban Water [online] Sustainable Water Management in Urban Space, http://www.urban-water.org/cms/. Accessed 6 May 2011.

Vickers 2001 'Handbook of water use and conservation', Waterplow Press Amherst, Massachusetts.

Walker, W. 2000, 'Entrapment in large technology systems: institutional commitment and power relations', *Research Policy*, vol. 29, pp. 833–846.

Water Services Association of Australia 2009 *Vision for a sustainable urban water future*. Position Paper No. 03. January 2009.

WHO 2006, *Guidelines for the safe use of wastewater, excreta and greywater, Volume 4: Excreta and Greywater use in Agriculture*.

Wickson, F., Carew, A.L. & Russell, A.W. 2006, 'Transdisciplinary research: characteristics, quandaries and quality', *Futures*, vol. 38, pp. 1046–1059.

World Bank 2009, *Commodity Price Data (Pink Sheet), Prospects for the Global Economy*, World Bank, Available: http://go.worldbank.org/5AT3JHWYU0.

WSRW 2007, *The phosphate exports*, Western Sahara Resource Watch, http://www.wsrw.org/index.php?cat=117&art=521.

Yarra Valley Water 2009, Yarra Valley Water Annual Report 2008–09. Available from http://www.yvw.com.au/yvw/Home/AboutUs/ReportsAndPublications/annual-reports.htm.

Young, O. 2002, *The Institutional Dimensions of Environmental Change: Fit, Interplay, and Scale*, MIT Press.

Chapter 5

The influence of water on urban energy use

Steven. J. Kenway[1] and P. Lant[2]

[1]*University of Queensland, Advanced Water Management Centre, Brisbane, Australia*

[2]*University of Queensland, School of Chemical Engineering, Brisbane, Australia*

ABSTRACT

This chapter provides an overview of the diverse and significant links between urban water and energy use. It is motivated by the rapidly increasing energy use in Australian urban water systems with resultant cost and environmental impacts. The chapter identifies opportunities for urban water management in Australia to contribute far more substantially to the reduction of water-related energy use. In particular, this includes energy use for water heating and cooling for residential, industrial, commercial and other purposes. In 2001, water-related energy use in California comprised 19 and 32% of total electrical and natural gas use respectively. Australia is likely to be of similar magnitude. Despite the significance, the connections are poorly understood and largely ignored. This is possibly because many of the influences are indirect, difficult to measure, change regularly and are outside the typical boundary of urban water responsibilities. Additionally, there is a current lack of an overall analytical structure within which to consider, let alone manage, the interconnections. The chapter broadly describes how the water-energy nexus connects to other sustainability issues together with the concept of urban metabolism which is perceived as critical to providing a structure for analysis. Finally, the chapter reflects on future profiles and implications in a future constrained by water and energy simultaneously.

Keywords: Water-related energy; urban metabolism; future cities

5.1 INTRODUCTION

Water service providers in Australia are facing rapid and substantial increases to energy consumption. The water sector has been active in reducing energy use and associated greenhouse gas emissions. However, these measures alone are unlikely to cope with the magnitude of forecast energy increases. A significant opportunity for the sector lies in harnessing the influence of water management on energy use indirectly.

In 2005–06, following unprecedented drought, a then record \$2.4 billion (AUD) was required for new water infrastructure for major Australian urban centres (Figure 5.1). Upgrading to climate-resilient supplies was part of the rationale articulated (WSAA 2008). In 2007–08, almost \$4 billion in capital was expended. This grew to over \$7 billion the following year. The forecast for 2009–10 is greater than \$14 billion (WSAA 2009).

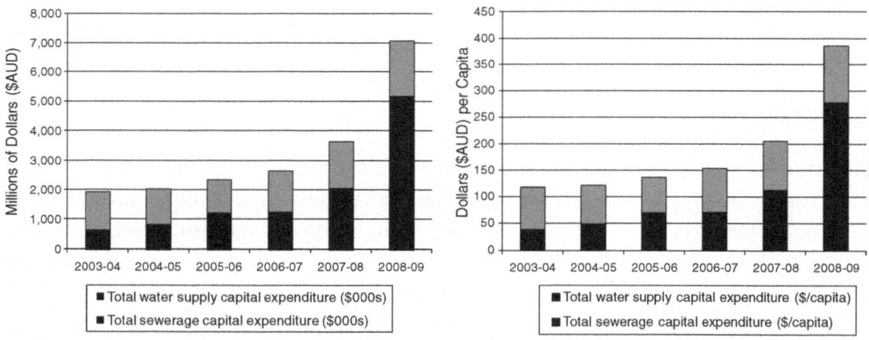

Figure 5.1 Capital expenditure trend (\$AUD) for water and wastewater in Australia (2002–03 to 2008–09), Data from: (National Water Commission 2010). Expenditure for all major cities in Australia with a collective population of 18.6 million people in 2008–09 representing 84.6% of the total Australian population

Most new sources of water are more energy-intensive because more extensive treatment or pumping is required per volume of delivered water. Direct energy use required to provide centralised water and wastewater services (e.g. by water utilities) in Australia, in 2030 is forecast to grow up to 130 to 200% above the 2006-07 levels (Figure 5.2). This assumes residential water consumption rates remain at 225 L/(capita $*$ d)(Kenway et al., 2008a) and a 25% population growth. If water consumption rates return to pre-drought consumption levels of 300 L/(capita $*$ d) residential use, and if the all new water is sourced from desalination, then energy use could grow 400% above 2006–07 levels.

However, the direct energy use necessary for water supply and sanitation would grow substantially more if climate change causes existing supplies to be reduced, even if water use remained constant. This is because water would need to be sourced from further away, from deeper groundwater sources, or of more marginal quality requiring more treatment. The Water Services Association of Australia have flagged the possibility of a 25% reduction in existing yields in some catchments (WSAA 2005). Consequently, under worst case scenarios (high water use, decreased existing yields, high population growth, increasing treatment standards, and sourcing all water from desalination) it is possible that energy demand for water use could grow 500% above 2006–07 levels.

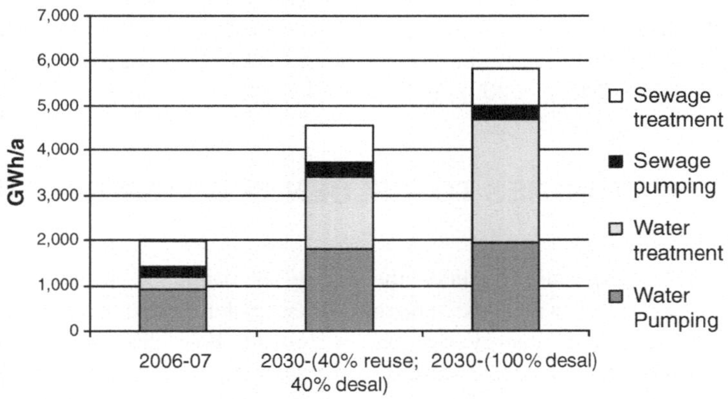

Figure 5.2 Energy use breakdown for Australian Urban Water Provision in 2006–07, and forecast to 2030 under two scenarios of water sources (Source (Kenway et al., 2008a))

In 2006–07 the majority of energy demand of the Australian Urban Water Sector was met through carbon-intensive coal-fired power (Kenway et al., 2008a). Purchase of Green Power (low emissions energy) is a current policy of many utilities to help offset energy use. While this is helpful, it does little to reduce the upward trend in energy-use. Purchase of green power also arguably creates a water system dependent on the price of green energy, and shifts credit for 'carbon-neutrality' from water providers to clean energy providers.

While agreement on global carbon reduction goals appears to have stalled, a number of governments (e.g. the United Kingdom, Japan, Sweeden, Norway, State of California) have committed to reducing national greenhouse gas emissions 80% of 1990 levels by 2050 (United Kingdom Government 2009). If such a five-fold reduction goal was to be applied to the Australian urban water sector, and if energy use did increase 200–500% above 2006–07 levels then there would be a 10 to 25-fold difference in the energy use trajectory, and pathway for desired emissions reduction. To address such a gap requires step-change, rather than incremental improvement.

The realisation is growing that managing water, energy, carbon and their connections, is vital to sustaining city and national prosperity. Hightower and Pierce (2008) articulate that in a water-scarce world, regulations, policies and infrastructure development must adapt to the reality that adequate freshwater supplies not only have health and social benefits, but economic benefits as well. Failure to do so will lead to stunted economic growth, inequitable development, and possibly regional conflicts. Low stream flows are already impacting hydro-power generation and operation of cooling towers in the United States (Pate et al., 2007). Similar effects are evident elsewhere. Drought in France in 2003 caused the loss of up to 15% of nuclear power generation capacity for five weeks and a loss of 20% of their hydropower capacity. Similarly, the 2007 drought across Eastern Australia raised concerns over water supplies and electric-power reliability (Marsh 2008). It is also likely to have led to an increase in energy prices.

5.2 OPPORTUNITIES TO INFLUENCE WATER-RELATED ENERGY

Water-related energy use includes the energy demands of using, as well as providing, water. If water use changes, then water-related energy use also changes. Water-related energy use can be broader than this and also include connections between water and wastewater strategies. For example, in the capture of carbon as useful methane from wastewater, but also in the energy value of nutrients lost through wastewater disposal. Finally, water management can also influence urban energy use through impacts on the natural evaporative cooling processes in the urban space.

Energy use by utilities is commonly considered. However, there are other uses that are often overlooked. These include energy used for the provision of decentralised water supplies, for example from rainwater tanks or bottled water. Similarly, additional energy may be required for pressurising water for delivery into high rise buildings. Energy is also used when water is used in the residential sector, largely through water heating however other uses such as swimming pool filters also consume energy. Industrial and commercial sectors also have large water-related energy use.

In 2006, 21 million Australians consumed 221 billion kWh of electrical energy (Index Mundi 2010). This is just over 10,000 GWh of end-use consumption per million people. Some 23,000 GWh (1,710 PJ) of primary energy (largely coal) was required to generate the electricity (ABARE 2008). In total, the Australian economy consumed just over 76,000 GWh per million people (5,770 PJ nationally). Considering a wide range of influences it is possible that alternative water provision and consumption practices affect around 1,200 GWh electrical and 2,600 GWh thermal (e.g. natural gas) energy use. Further clarification and quantification of these numbers is provided in a supporting paper (Kenway et al., in submission).

Australian cities offer significant scope for addressing the problem of rising energy use. Australia is already a heavily urbanised nation with over 80% of its population living in cities. This urban concentration will increase. Around 70% of Australia's projected population growth, from 21 million in 2007 to around 30 million, by 2030 (ABS 2009) is expected to occur in its capital cities. Much of the remaining growth will be in other major urban areas. Most of the economic value of Australia is either created in, or flows through, cities. Urban design represents a chance to reduce water-related energy use.

5.2.1 Energy use in the provision of water

While water and wastewater service providers are large individual energy consumers, the direct energy end use by utilities in Australia is less than 0.5% of total per-capita primary energy use (Kenway et al., 2008a). In California, where water is transported further, provision of urban water pumping and treatment consumes around 4% of State-wide electricity (Klein et al., 2005).

Average energy demands for water and wastewater provision in Australia in 2006–07 were approximately 0.8 and 0.7 kWh/m^3 respectively. Improved operations (e.g. water loss minimisation) or technologies (e.g. variable speed pumps) can help minimise the energy requirements of utilities. Local conditions can dramatically influence the energy needs for water or wastewater service provision. For example, in 2006–07 water ranged from 0.09 to 1.90 kWh/m^3 (Kenway et al., 2008a). The low range was for Melbourne where high quality water from protected elevated catchments drains by gravity to the city. The high end was for for Adelaide, where significant transport of lower-quality water was required. Energy for wastewater treatment and transport over the same period was 0.045–1.13 kWh/m^3. Energy self-sufficient wastewater treatment processes are not only possible, but are in operation to various degrees, largely through methane capture after anaerobic digestion of sludges.

Water policy can also affect the energy demands of water provided through decentralised systems. For example, over 150,000 rainwater tanks were installed in Brisbane between 2005 and August 2007 following the introduction of government rebates (Gardner 2007). The energy-intensity of rainwater tanks can range from 0.1 kWh/ m^3 for low pressure systems (Cunio and Sproul 2009) to over 5 kWh/ m^3 for example if UV treatment of water is required (Hood et al., 2010).

Bottled water use is likely to be affected by the quality, availability and pricing of the centralised water supply. Bottled water requires around 1000 times the energy-intensity of centralised water supplies (1500–2500 kWh/m^3, (Gleick and Cooley 2009). If bottled water use comprised 0.1% of total urban water use in Australia, this would equal the total energy used to provide centralised water in 2006–07. Strategies which promote use of centralised water over bottled water are likely to be large energy savers.

Finally, the shape and nature of our cities can also influence the energy demands for water itself (Kenway et al., 2008b), or water pumping. Each six stories of elevation adds around 0.15 kWh (Cheng 2002) for each cubic meter of water delivered.

5.2.2 Energy use associated with the use of water

Urban water management indirectly influences much larger amounts of energy, particularly in the use of water. In Australia, residential water heating alone accounts for approximately six to ten times more energy use than the energy use of all water utilities (Kenway et al., 2008a). Water-related energy use in households is substantial and includes heating water for shower, bath and tap use as well as clothes washing and dishwashing. In cold cities with ambient temperatures of 5 degrees Celsius, water heating comprises 97% of the energy use of residential water provision and use (Arpke and Hutzler 2006). However, even if ambient cold water temperature is 18 degrees, hot water energy use remains by far the dominant fraction of total energy use associated with the residential water cycle (Flower 2009).

Many actions are available to help customers reduce hot water use and save significant quantities of energy. Information programs, water-pricing structures and technology rebates (e.g. water-efficient clothes washers) are all examples of approaches that can be used to encourage customers to conserve hot water use.

Urban water management and policy can influence energy use by utilities as well as residential, commercial and industrial water users. Many factors however are likely to constrain the water sector, urban planners or governments in general from taking wider action in targeting indirect energy impacts. Why should they take responsibility for impacts outside their jurisdiction? Who gets the credit if energy is saved? Who pays? Who benefits? Don't customers have the right to decide how much water and energy they each use?

As most water businesses are remunerated in accordance with the volume of water sold, strategies which reduce the volume of water can impact on profitability, unless alternative incentives can be provided. This chapter contends that a water business could credibly argue that supporting wider and more significant urban systems energy use, could be more effective than focusing solely on their own energy use. Finding appropriate incentives and regulatory and governance structures to encourage this would be no small task.

Other energy use is also associated with residential water use. For example, around 2500 kWh/household/annum is used for filtering pool water. Low energy filtrations systems could help reduce this use. Increased use of public pools over private pools could also potentially reduce energy use. Around 135 kWh/household is used for boiling water in kettles (DEWHA 2008). Approximately one third of this is estimated as wasted through overfilling.

In California, Klein et al. (2005) note that the industrial and commercial sectors have influence on even more energy use than the residential sector. Typical uses of water that consume substantial energy are heating, cooling, filtering and air-conditioning. Technologies which use steam (e.g. for cleaning and cooking) are particularly significant in the water-energy connection. Saving small amounts of steam use and waste discharge (e.g. through altering blow-down criteria) can save substantial amounts of energy (deMonsabert and Liner 1998). Energy use associated with the industrial use of water represents a high-priority for analysis identified by several authors (Kenway et al., Accepted). It is particularly warranted considering the high magnitude and uncertainty of existing estimates.

5.2.3 Energy associated with the nutrient cycle

Other opportunities to influence energy use exist in the nutrient cycle. Water and wastewater policy influences the flow of nitrogen, phosphorus, water and food products in cities. Each tonne of nitrogen lost from our wastewater system requires a similar quantity to be resynthesised or imported for use in agriculture. Each tonne of food production displaced from our cities requires food to be imported back into cities, often at far higher total energy and water cost.

The pioneering German Chemist Justus Liebig wrote to the British Prime Minister in 1840 with concern that the proposed introduction of flushing toilets (water closets) would remove nutrients into the sea and that this would lead to soil nutrient impoverishment and consequential plant productivity decline. Nonetheless, the sewage and disposal option was instigated at the expense of the traditional reuse systems. This has subsequently led to the deterioration of receiving waters, declining soil fertility levels and a rapid dependence on high-energy synthesis of manufactured fertilisers.

Barles (2007) used urban metabolism to identify that approximately 40% of the nitrogen intake of Paris in 1914 was sourced from reuse of the nutrients in human and animal waste. This high recycling rate was enable partially because the relatively dry (unsewered) waste was easier to compost and collect. The fraction of Nitrogen recycled in Paris today is likely to be far less. As a result, the energy intensity of the food consumption of Paris now is likely to be far higher.

The role that the nutrient cycle plays in water and energy is only beginning to emerge. In a future constrained by water and energy this will become increasingly evident. An analysis of the water and energy futures of Melbourne revealed that the total water necessary to sustain Melbourne includes not only around 350 GL/a of direct water use (largely for consumption within residential, commercial and industrial sectors), but also approximately 60 GL/a in electricity and 1,400 GL/a in food (Kenway et al., 2008b). Melbourne also relied on approximately 1,500 GL/a water (its pro-rata share of national water use) to provide food exports. This means the indirect reliance of Melbourne on water (largely to drive the nutrient cycle and food production for cities) is nearly 10 times its reliance on water directly. Our decision-makers may need to make choices between limiting water use for (1) direct consumption (2) energy production (3) food consumption or (4) food or product exports. Alternatively, it would be necessary to rapidly develop radically new technologies to lower the energy and water intensity of the existing economy.

5.2.4 Urban heat island effect

There are a range of other more distant and difficult to characterise connections between water and energy in cities. For example, water management can also influence urban temperatures. In some cities, temperatures at the tree canopy layer can be as much as 10°C higher than the surrounding rural areas (Kennedy et al., 2007). Loss of water and cooling effect of evapo-transpiration from the urban area, and release of waste heat from energy use are the two primary causes. Such temperature differences are certain to affect energy use such as air conditioning demands.

5.3 THE WATER-ENERGY NEXUS IN THE BIGGER PICTURE OF URBAN METABOLISM

Urban Metabolism is a conceptual model to facilitate the description and analysis of the flows of materials, including water and energy, within cities (e.g. Wolman 1965; Newman 1999; Decker et al., 2000). At its simplest, urban metabolism considers the mass balances of all materials of urban systems (Sahely et al., 2003). At its most complex, the concept draws on the rich metaphor which considers cities literally as organisms or super-organisms (Wolman 1965; Decker et al., 2000).

The idea of urban metabolism was first put forward with a core intent of addressing contemporary urban resource issues nearly 50 years ago by Abel Wolman (1965). However, the roots of the theory go back to the mid 19th Century. At the time, the divide between the social sciences and humanities on the one hand and the natural sciences on the other, was not as wide as in the current time (Fischer-Kowalski 1998). In the 1850's metabolism was considered an exchange of matter between an organism and its environment, rather than a cellular biochemical conversion (Fischer-Kowalski 1998).

In his pioneering article, Wolman (1965), a water planner, was motivated to use metabolism theory to simultaneously address the then clear problem of 'shortages of water and the pollution of water and air', together with a call for 'public economic decisions'. He used the urban metabolism concept to demonstrate that there is plenty of water, but providing water without creating problems elsewhere requires both foresight and a willingness to pay.

Wolman defined urban metabolism as all the materials and commodities needed to sustain city inhabitants at home, at work and at play including the construction materials needed to build the city. In a hypothetical city of 1 million people, Wolman noted the dominance of water in the material needs. Water comprised over 98% of the urban mass balance. His city relied on 625,000 t/day

(0.6t/(capita $*$ d)) of water which entered 'silently and unseen'. The same city produced around 500,000 t/day (0.5 t/ (capita $*$ d) of sewage. Total fuel consumption was approximately 9,500 t/d, and food input around 2,000 t/d. Wolman (1965) also commented on the complexity of the urban energy cycle by pointing out its distinct differences to the urban water cycle, such as the lack of a centralised waste collection system that sewers provides.

More recently, Sahely et al (2003) indicated that the concept of the ecological footprint (Wackernagel and Rees 1996) is a derivative of the metabolism model. Ecological footprint is a spatial land measure of land necessary to support a city. Sahely points out however that the ecological footprint concept does little to characterise flows through a city and to identify areas of particular concern as intended by Wolman in his original model.

While urban metabolism considers the rate of flow of energy, water, and all other materials through cities, the water-energy nexus is specifically focussed on the inter-linkages. This chapter contends that understanding the water-energy nexus alone, would be like trying to understand human health solely by observing sub-systems of the body (for example the cardio-vascular system), without any knowledge of what was happening to the body in question itself. Does it have a good diet? Is it healthy or sick? Is it running or sleeping? Considering urban metabolism simultaneously with the water-energy nexus provides critical context to our discovery and understanding of relevance of the connections.

5.3.1 Recent urban metabolism analysis

In the 40 years since Wolman published his pioneering article, only a limited number of metabolism studies have been conducted. Almost all have focussed on broad inputs and outputs of individual cities including Vienna, Tianjin, Sydney, London, Hong Kong, Cape Town, and Toronto (For references see Decker et al., 2000; Kennedy et al., 2007). In most cases, the research has been aimed at understanding the throughput of cities. Knowledge to help reduce the impact of resource requirements and discharges of cities has been the motivator.

The detailed methodology of urban metabolism or materials flow analysis are outlined in Baccini and Brunner (1991). The major steps in analysis of urban metabolism analysis include: (1) definition of goals and research questions, (2) system description, (3) data acquisition, (4) material balances, and (5) interpretation (Sahely et al., 2003).

Decker (2000) systematically compared the metabolism of 20 megacities with regard to their food, fuel, water and air cycles. He noted that 'megacities are somewhat independent of their immediate environment for food, fuel, and aggregate inputs, but all are constrained by their regional environment for supplying water and absorbing wastes'. He reinforces Wolman's work that the flux of water stands out as the dominant material flow through megacities, and concludes that increasing urban water demand will require passive inputs of water to be treated as a resource, necessitating changes in water management infrastructure. Decker notes that 'atmospheric pathways surface as the most important for understanding impacts of megacities on neighbouring ecosystems and at the earth system level'. He also notes the paucity of information on food flows despite their great significance for the nitrogen cycle and solid waste management.

Decker (2000) concludes that little effort to date has been invested in comparing the growth of megacities. He suggests that understanding the energy and materials processes of urban systems is an imperative for addressing the social, environmental and energy challenges of the next century. As a priority he indicates that correlating input and output variables (e.g. water, wastewater, energy, greenhouse gas emissions) of urban systems is necessary to enable material and energy flux predictions for arbitrary cities. It would also provide critical information about energy efficiency, material cycling, waste management, and infrastructure architecture in urban systems.

Kennedy (2007) considered the changing metabolism of eight cities since 1965. He analysed four fundamental flows or cycles: water, materials, energy, and nutrients. Kennedy notes that the data suggest that the metabolism of cities is increasing. Water and wastewater flows were typically greater for studies in the 1990s than those in the early 1970s. He concluded that only one city, Toronto, improved metabolic efficiency, possibly as a result of changes to industrial processes. He also indicated that the paucity of published data on cities makes trend analysis impossible. He noted that resource accounting and management is typically undertaken at national levels and that this practice may constrain understanding of the urban driving processes. He suggests that the difference between the input from and output to surface waters may be as important as the sheer volume of supply due to issues of accumulation (for example of groundwater and associated infrastructure damage) or exhaustion. Kennedy notes that accumulation processes, such as ground water flux or heat storages in roof tops and pavements, should be understood so that resources can be used appropriately.

Kennedy (2007) notes that urban policy makers should be encouraged to understand the metabolism of their cities so they know if they are using water, energy, materials, and nutrients efficiently, and how this efficiency compares to that of other cities. They must consider to what extent their nearest resources are close to exhaustion. If necessary, urban metabolism analysis could guide appropriate strategies to slow exploitation.

5.3.2 Recent application of the metabolism model

The urban metabolism model has been largely only used in the academic domain. Very few examples of its application into strategy, policy, regulation or serious evaluation can be found. Newman (1999) demonstrated the relevance of urban metabolism in framing city performance indicators.

Another example of using metabolism to derive urban performance indicators was undertaken by Kenway et al (Kenway et al., Accepted, in revision). Water mass-balance analysis of Australian cities demonstrated that large flows passed though drought-stressed cities unaccounted for and under-utilised. In 2004-05, Sydney, Melbourne, South East Queensland and Perth varied from 26 to 86% in their potential to meet total water use by wastewater recycling. The cities also varied from 47% to 104% to meet water use by recycling urban stormwater. The work demonstrates how mass balance can derive quantitative indicators of whole of urban system performance and validate the consistency of data flows through cities.

Lack of unifying theoretical framework and methodological coherence contributes to lack of a consistent set of operational tools in metabolism research, even though the concept has emerged as one of the most powerful paradigms and interdisciplinary concept for the empirical analysis of the society-nature interaction (Fischer-Kowalski and Huttler 1998). Satterthwaite (2008) suggests that most large cities have three or four different boundaries: (1) the core; (2) the contiguous built up area; (3) the metropolitan area (4) extended planning region. He points out that our current loose definition of cities leads to great difficulty in comparing even basic parameters including population which can vary by several million persons depending on definitions adopted.

A critical step in the analysis of a system (for example a city) is the identification of the boundary (Sterman 1991). Identification of a consistent urban analysis boundary could improve the value of the various environmental analysis methods currently used for options comparison. For example Environmental Impact Assessments, Life Cycle Assessments, and planning strategies are currently undertaken against a wide range of boundaries, rather than considering a consistent boundary. Adoption of different boundaries could lead to inclusion or exclusion of very different sets of linkages and consequently very different overall conclusions. For example, if the energy implications of the residential 'urban water cycle' are considered with, or without, the energy implications of end-use, up to 97% of the energy implication could be missed.

A consistent boundary may also help to identify which factors are assumed to be causal to, and which a consequence of, our actions. Without knowing how the cause-and-effect implications of water management or other urban strategies play out against a consistent reference boundary, it is impossible to know which factors are critical influencers or driving forces in the analysis. Because water represents the vast mass of material moving through cities, it is possible that water will define a clear urban system boundary. Further, a consistent mass-balance defined boundary could give us a boundary against which optimisation of resource flows (converting water, energy and nutrients into human well being and human well-being) could be achieved. Without such a boundary, we will have to contend with a faint hope that optimising (and strategising) around sub-systems, such as water, energy, transport, will result in an overall improvement in the efficiency of urban systems performance.

5.4 WATER AND ENERGY INTEGRATION FUTURES

Many developments are necessary if our understanding and management of the water-energy nexus are to improve. Many of the steps are iterative. Improvement is required in our analysis methods. Better quality data is required to inform the analysis. And before the data can be collected or assembled, there is a need for wider understanding of the significance of the issue. The first three rows of Table 5.1 describe the history and future of these three elements. Historically the relevant data sets have been widely scattered, issues were considered in isolation and, not surprisingly, there have been few consistent methods applied to assess water-energy connections. In future this will be much more connected, comprehensive and standardised.

Improved analysis and data will help us derive indicators of urban performance which currently ignore trade-offs between water and energy. Future performance indicators of urban systems could be expected to simultaneously target water and energy consumption trends. Based on improved targets our monitoring and data compilation programs will improve, as will the reporting and the understanding of the issues.

Ultimately co-ordinated water and energy regulation is necessary to drive planning and behaviour systematically in a direction that reduces the metabolic rate of our cites. Some nations are already planning to integrate aspects of water and energy legislation, and this trend can be expected to accelerate.

Reduction in the metabolism (total material and energy throughput) of our cities is required if the total impact of our cities are to be systematically reduced. It is also necessary if we wish to know if our cities are indeed on the path to more sustainable forms, or if they are simply shifting problems from one domain to another. Input from the academic, government, and industry sectors will be necessary to progress this.

The World Business Council suggests 'a global trend towards increased corporate accountability for products use. In the coming years, it is likely that companies will increasingly report across the value chain including wider downstream (consumer related) impact of products and services' (Heemskerk et al., 2002). The global reporting initiative (GRI 2005) sector supplement for public agencies flags a similar direction, encouraging public entities to report on the impacts associated with the use of their products, not just of their operation. However, with regard to energy implications, for the majority, the current content is focussed solely on direct utility-level energy use. Some utilities (eg. Sydney Water, Yarra Valley Water) have undertaken some reporting on the influence of their policies on the energy use of their customers.

Table 5.1 The emerging landscape of water-energy linkages

Element	Where we were (1990)	Where we are now (2010)	Where we are going (2030)
Understanding and management.	Issues considered and managed in isolation.	Awareness of the connection however management is fragmented and without any consideration of or consistency, particularly for indirect connections.	Comprehensive simultaneous consideration and management of the linkages between water and energy.
Analysis methods.	No standard methodologies.	Inconsistent methods focussed on useful but narrow components of the problem and/or solution (e.g. water sources, or wastewater technologies).	Standardised approaches unifying analysis and identifying whether solutions are addressing the root cause of total material and energy flux's through cities.
Data necessary for analysing water-energy connections and overall urban performance.	Widely scattered.	Improved availability but highly patchy.	Coordinated and inter-locking to enable diverse analysis and management of water and energy simultaneously.
Urban resource strategies.	Separate water and energy strategies developed for cities. Lack of awareness leads to 'unplanned' trading of water, energy and nutrient outcomes and impacts.	Awareness that water and energy management need to be integrated. Emerging awareness that the nutrient cycle is critical.	Fully integrated strategies addressing water, energy and nutrient requirements of cities.

(*Continued*)

Table 5.1 (*Continued*)

Element	Where we were (1990)	Where we are now (2010)	Where we are going (2030)
Urban systems performance indicators for water and energy flows.	Lack of comment or consideration of the issue.	Fragmented and unquantified. Emerging evidence of the need.	Quantified performance indicators which simultaneously consider water and energy.
Reporting and benchmarking of urban performance with regard to water, energy and nutrient efficiency.	Identified as desirable however not attempted.	Attempted however insufficient data sets.	Standardised and used to evaluate the efficacy of urban design and management together with indicators of human well being (including urban economic performance) and ecosystem health.
Regulation of water and energy issues.	Largely separate.	Discussion about and proposals for integration (e.g. Water and Energy Integration Bill in the USA). Separate water and energy utilities.	Legislative and governance review to ensure water and energy issues and efficiencies simultaneously through planning and regulatory mechanisms. Incorporated into legislative targets and planning processes (e.g. regional planning and building approval processes).

5.5 CONCLUSIONS, IMPLICATIONS AND RECOMMENDATIONS

Water management can influence urban energy use substantially. Water could be a lever to reduce urban energy use. Conservation of hot water in the residential and industrial sectors, or provision of clean energy sources for water heating are strategies which could significantly reduce water-related energy use, and greenhouse gas emissions.

Many of the links between water and energy are not well understood or considered, for example in planning decisions. Part of the reason is because the management of our city-systems has been reduced to its components of 'water', 'energy' and 'transport' and the 'food' system pretty much looks after itself. To date, little effort has been expanded to optimise the city overall, for example, with regard to its water and energy performance. Compartmentalising the problem in this way means that the performance indicators that guide decision making in each sector are narrowly focused and give no consideration to how the city as a whole functions and performs. There appears to be minimal direct incentives for the water sector to help mitigate total urban energy use.

Lack of methodological standardisation in the analysis of cities creates difficulties for trans-city comparisons. This shortcoming is likely to be due to the slow adoption of an appropriate unifying theoretical framework within which to consider the water-energy nexus. At best, the lack of method and standards makes it difficult to compare results. At worst it may raise questions about the validity of previous studies.

The concept of urban metabolism provides a conceptual and theoretical basis against which investigation of the water-energy nexus in cities could have far greater rigour and quantification. While the concept has been largely constrained to the academic domain to-date, there is a real need to take the concept into planning, policy, monitoring and regulatory applications. A challenge to be overcome is the fragmentation of the many necessary data sets. Wider use, understanding, and reporting of urban metabolism are an essential future of water and energy in cities. It will help us understand whether a particular option really does reduce energy use, or whether it simply shifts it from one area to another – or translates a water problem into an energy problem, or vice versa.

A suggested starting point for managing 'water-related energy use' could through appropriate reporting. This is often the first step in improving the knowledge on which subsequent actions are based. Sustainability, triple bottom line or corporate social responsibility reports are already produced by many utilities annually. Moving to report on a wider sphere of impacts is a logical next step. Further recommendations are provided in (Kenway et al., 2008a).

In pursuing goals of water and energy use reduction, it is likely to be highly beneficial for the water and energy sectors to collaborate. Co-optimising systems for water and energy could lead to very different outcomes than optimising for either one of these elements independently.

Local conditions have a critical influence and must be considered in any specific recommendations. It is important that results from one area or city are not translated, without validation, to another area.

As the challenges of climate change continue to unfold, and mitigation targets progressively aim to reduce emissions further, taking a wider perspective will be necessary for the water sector to contribute proportionally with other sectors in achieving targeted reductions in energy use and associated greenhouse gas reductions. Understanding the wide, diverse, and ever-changing connections between water and energy will help find strategies which enable this in the most cost-effective manner.

ACKNOWLEDGEMENTS

The support of the Urban Water Security Research Alliance in preparation of this paper is acknowledged. The helpful review of Dr Korneel Rabaey and Joe Lane are appreciated. The primary author acknowledges the support of eawag, the Swiss Federal Institute of aquatic research where this paper was finalised.

REFERENCES

ABARE (2008). Australian Energy Statistics: Australian Energy Update 2008.

ABS (2009). 3101.0 Australian Demographic Statistics September 2009. A. B. o. Statistics. Canberra, Australian Bureau of Statistics.

Arpke, A. and Hutzler, N. (2006). Domestic Water Use in the United States. A Life-Cycle Approach. *Journal of Industrial Ecology*, **10**(1–2), 15.

Baccini, P. and Brunner, P. H. (1991). *Metabolism of the anthroposphere*. Berlin, Springer Verlag.

Barles, S. (2007). Urban metabolism and river systems: An historical perspective-Paris and the Seine, 1790–1970. *Hydrology and Earth System Sciences*, **11**(6), 13.

Cheng, C. (2002). Study of the Inter-Relationship Between Water Use and Energy Conservation for a Building *Energy and Buildings*, **34**, 6.

Cunio, L. N. and Sproul, A. B. (2009). Low Energy Pumping Systems for Rainwater Tanks. *Solar09, the 47th ANZES Annual Conference*.

Decker, E. H., Elliott, S., Smith, F. A., Blake, D. R. and Rowland, F. S. (2000). Energy and material flow through the urban ecosystem. *Annual Review of Energy and the Environment*, **25**, 685-740.

deMonsabert, S. and Liner, B. (1998). Integrated water and energy conservation modelling. *Journal of Energy Engineering*, (April), 18.

DEWHA (2008). *Energy Use in the Australian Residential Sector 1986–2020*. Canberra, Commonwealth of Australia. Department of the Environment, Water, Heritage and the Arts.

Fischer-Kowalski, M. (1998). Society's Metabolism: The Intellectual History of Materials Flow Analysis, Part I, 1860-1970. *Journal of Industrial Ecology*, **2**(1), 61–78.

Fischer-Kowalski, M. and Huttler, W. (1998). Society's Metabolism: The Intellectual History of Materials Flow Analysis, Part II, 1970–1998. *Journal of Industrial Ecology*, **2**(4), 30.

Flower, D. J. M. (2009). An Integrated Approach to Modelling Urban Water Systems. *Department of Civil Engineering*. Melbourne, Monash University. **Doctor of Philosophy**, 373.

Gardner, T. (2007). Rainwater Tank Rebates Paid to August 2007, Science Seminar Presentation, Indooroopilly, 24th October. S. J. Kenway. Brisbane, Department of Natural Resources and Water, Queensland.

Gleick, P. H. and Cooley, H. S. (2009). Energy implications of bottled water. *Environmental Research Letters*, **4**, 6.

GRI (2005). Sector Supplement for Public Agencies: Pilot Version 1.0. Boston, Global Reporting Initiative.

Heemskerk, B., Pistorio, P. and Scicluna, M. (2002). Sustainable Development Reporting: Striking the Balance. Geneva, World Business Council for Sustainable Development.

Hightower, M. and Pierce, S. A. (2008). The energy challenge. *Nature*, **452**(20 March).

Hood, B., Gardner, T., Barton, R., Gardiner, R., Beal, C., Hyde, R. and Walton, C. (2010). The Ecovillage at Currumbin - A Model for Decentralised Development. *OZWATER 2010*, Brisbane, Australian Water Association.

Index Mundi (2010) Australia Electricity - Consumption.

Kennedy, C., Cuddihy, J. and Engel-Yan, J. (2007). The changing metabolism of cities. *Journal of Industrial Ecology*, **11**(2), 43–59.

Kenway, S. J., Gregory, A. and McMahon, J. (Accepted, in revision). Urban Water Mass Balance Analysis *Journal of Industrial Ecology*.

Kenway, S. J., Lant, P. and Priestley, A. (in submission). Quantifying the links between water and energy in cities. *Journal of Water and Climate*.

Kenway, S. J., Lant, P., Priestley, A. and Daniels, P. (Accepted). The connection between water and energy in cities - a review. *Water Science and Technology*.

Kenway, S. J., Priestley, A., Cook, S., Seo, S., Inman, M. and Gregory, A. (2008a). *Energy use in the Provision and Consumption of Urban Water in Australia and New Zealand*, CSIRO and Water Services Association of Australia.

Kenway, S. J., Turner, G., Cook, S. and Baynes, T. (2008b). *Water-energy futures for Melbourne: the effect of water strategies, water use and urban form*. Canberra, CSIRO.

Klein, G., Krebs, M., Hall, V., O'Brien, T. and Blevins, B. (2005). California's Water-Energy Relationship - Final Staff Report, California Energy Commission.

Marsh, D. (2008). The water-energy nexus: A Comprehensive analysis in the context of New South Wales. *Faculty of Engineering and Information Technology*. Sydney, University of Technolgy. **Doctor of Philosophy (Engineering)**, 370.

National Water Commission (2010). National Performance Report 2008-09 - urban water utilties. Part C.

Newman, P. W. G. (1999). Sustainability and cities: extending the metabolism model. *Landscape and Urban Planning*, **44**(4), 219–226.

Pate, R., Hightower, M., Cameron, C. and Einfeld, W. (2007). Overview of energy-water interdependencies and the emerging water demands of energy resources. Albuquerque, New Mexico Sandia National Laboratories.

Sahely, H. R., Dudding, S. and Kennedy, C. A. (2003). Estimating the urban metabolism of Canadian cities: Greater Toronto Area case study. *Canadian Journal of Civil Engineering*, **30**(2), 468–483.

Satterthwaite, C. (2008). Cities' contributions to global warming: notes on the alocation of greenhouse gas emissions. *Environment and Urbanization*, **20**(2), 539–549.

Sterman, J. (1991). A skeptic's guide to computer models. Massachusetts Institute of Technology.

United Kingdom Government (2009). Climate Change Act Impact Assessment. D. o. E. a. C. Change, 126.

Wackernagel, M. and Rees, W. (1996). *Our Ecological Footprint: Reducing Human Impact on the Earth*. Gabriola Island, New Society Publishers.

Wolman, A. (1965). The Metabolism of Cities. *Scientific American* **213**, 179–190.

WSAA (2005). Testing the Water - Urban Water in Our Growing Cities: the Risks, Challenges, Innovation and Planning Melbourne, Water Services Association of Australia.

WSAA (2008). *WSAA Report Card 2007-2008. Performance of the Australian Urban Water Inustry and projections for the future.*, Water Services Association of Australia.

WSAA (2009). *WSAA Report Card 2008-2009. Performance of the Australian Urban Water Industry and projections for the future.*

Chapter 6

A framework for developing sustainable water utilities in the coming decades

Peter D Binney

Director of Sustainability, Merrick & Company

ABSTRACT

Urban water managers in both developed and developing settings can consider best management practices and case studies from peer organizations when developing organizational and programmatic approaches to addressing their future needs. Many of the challenges faced by those managers are common to other locations and while they will consider solutions appropriate to local conditions, there are a set of key success factors and guiding principles that are applicable to most settings. Sustainability principles can provide a valuable framework for not only increasing eco-efficiency but also the organizational capacity to collaboratively identify more effective and supportable approaches to meeting the growing challenges of reliable water services to urban areas around the world.

Keywords: Urban water utility management; benchmarking; sustainability; asset management; water resources planning

6.1 INTRODUCTION

Water utilities around the world are facing growing challenges in providing safe and adequate water services to expanding service populations that are increasingly concentrating in urban areas. Requirements for additional water resources and services are placing greater demands on constrained and more variable water resources that must also serve energy, industrial, agricultural, social and environmental needs. Governance and utility operational models, funding mechanisms, regulatory approaches and public policies that satisfactorily served the needs of the 19th and 20th century are increasingly seen as inadequate for meeting 21st century challenges. The large costs and political challenges in building effective water infrastructure solutions now often require significant changes to the traditional approaches of building and operating single purpose water projects.

A growing trend among water utilities who are effectively responding to these challenges includes a number of adaptations that are practical applications of sustainability principles. Those changes incorporate environmental protection within a broader approach to financial stewardship and community involvement. Water utilities that are proactively and successfully developing operational and capital programs in today's complex regulatory and policy governance model also exhibit a number of key success factors. Case studies are described for consideration by other water utilities who are responding to the challenges of growing demands, water scarcity and conflict, climate change, aging assets, fiscal constraints and evolving governance.

6.2 CHALLENGES FACING WATER PROVIDERS

The primary mandate for water utilities is to provide safe, adequate and reliable water services for current and future needs.

There is, however, an increasing infrastructure and service gap in many developed nations that is testing the adequacy of current conventions and operating principles. That gap is mirroring the broader deficit in water services in developing countries which are attempting to meet the ambitious agenda of the Millennium Development Goals for improving the human condition by 2015 (United Nations, 2000). In the United States alone, it is estimated that the 20-year capital needs for water and sanitation could exceed $750 billion which is far in excess of the available funding sources at current pricing levels (US Environmental Protection Agency, 2002 and 2009). World water infrastructure needs could exceed one trillion dollars annually (Booz, Allen, Hamilton, 2006).

The financial demands to meet the needs of aging assets and deferred or planned infrastructure are impacting many water providers' plans for maintaining and increasing the service levels of their water systems. This could more frequently affect regional economies as seen in California in 2009 when it was estimated that agricultural revenue losses exceeding $3 billion were caused by water shortages. During the drought of 2007 in the south-eastern United States, water shortages were called the 'biggest and most imminent economic threat to the region' by the president of the Metro Atlanta Chamber of Commerce. Even though the states of Alabama, Georgia and Florida had been engaged in more than 20 years of litigation and negotiation over river and reservoir operations, the matters were still unresolved. This resulted in impacts to power production, industrial and manufacturing sectors, and agriculture, and pitted human needs against ecosystems protected by federal environmental protection laws. The chief federal official responsible for operations on the river system noted that the experience indicated the need for a major reconsideration of how freshwater resources are viewed and managed. Australia has been experiencing similar issues with the drought effects of the early 21st century. Water providers are now rapidly responding with a comprehensive range of integrated and sustainable water management programs. This empirical evidence of the effects of drought and water shortages on the adequacy of water supply systems is just one catalyst towards the evolving approaches to water management as described in this chapter.

Several other contributing factors that are requiring substantial responses include the rapid growth of urban populations and the increasing relative importance of urban-centric economies to national welfare. They also include climate change, environmental considerations as well as increasing competition between various water-using sectors and regions for scarce resources. The evolution of water governance institutions and regulations can, at times, be contradictory and ineffectual in delivering sustainable solutions.

Globally, we consume water at a rate that is doubling every 20 years and by 2025, it is estimated by the United Nations that the demand for fresh water will rise by nearly 60 percent more than is currently available. The world population is projected to increase from about 7 billion people in 2009 to nearly 10 billion by 2050. Urban populations will increase from today's 3.3 billion to more than 6.4 billion concentrated in 27 megacities by 2050, up from 19 megacities today. By comparison, urban populations increased from 220 million to 2.8 billion in the entire 20th century. More than 50 percent of the world's population will live in cities of less than 500,000 people. Many of those smaller urban areas do not currently exist but will develop as extended economic units of existing major metropolitan areas. It is unclear whether the utility needs including transportation, energy and water services of these newly developing areas will be incremental extensions, regionally integrated networks or independent systems. Because of the common natural resource base and watersheds that will serve the region, the approach to meeting the service needs of the newly developing areas will be a major factor in how future water systems will be constructed and operated and whether they will be sustainable.

Climate change and climate variability are projected to have a major effect on the adequacy and reliability of source waters for water users and ecosystems. In a recent study (National Resources Defense Council, 2010), it was projected that more than one third of the counties in the United States would face higher risk of water shortages by mid-century. When climate change effects are considered, nearly one half of those being considerd are at extremely high risk of water shortages. As seen in Figure 6.1, those at-risk counties are concentrated in the south-western and south-eastern states where proportionately higher population growth is expected. Those effects would also be seen in the irrigation-dominated Mid-western states that rely on groundwater resources from the depleting Ogallala Aquifers. Figure 6.2 is a retrospective view of inflows to the Water Corporation's reservoir systems in Perth, Australia. That trend clearly illustrates the long term drop in rainfall derived inflows to reservoirs. This is a key driver behind the development of the comprehensive *Water Forever* program (Water Corporation, 2009) that is now guiding that water provider's actions in developing a sustainable water program over the next 50 years. Other major water utilities in Australia including Melbourne Water and Sydney Water have developed similar integrated and sustainable water management plans that can serve as effective examples of industry leading water provider practices. Elements of these comprehensive programs are described later in this chapter and are instructive in their breadth of demand management and supply side approaches.

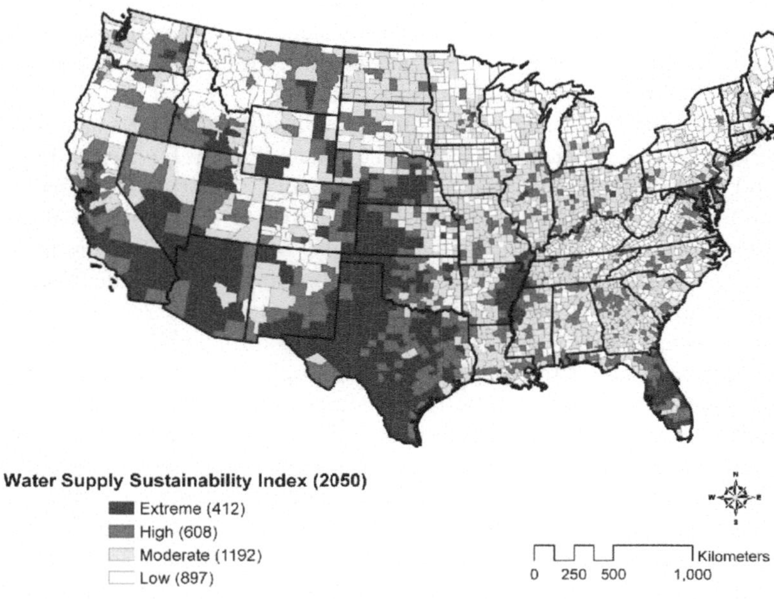

Figure 6.1 Projected Water Supply Sustainability in the United States in 2050. *Source*: Evaluating Sustainability of Projected Water Demands under Future Climate Change Scenarios, July 2010, Natural Resources Defense Council *(reprinted with permission)*

The growth in population and consumption patterns is also driving higher demands of water for agriculture and energy production. Agriculture is the largest consumer of fresh water resources (in Colorado, more than 85% of water consumed is for irrigation uses). Water for cooling thermoelectric power generation in the United States withdraws about one half of all daily diversions (US Geological Survey, 2009). There are also major allocations of water for recreational, ecosystem and environmental uses including instream flow protection programs. Even though water needs for public and domestic use typically account for less than 10 percent of all water withdrawals, they must be considered with other consumptive and non-consumptive uses in a watershed. It has been observed that a 'blue-green' revolution will be needed to increase the relative productivity of agricutlural water practices if the world's food production is to double by 2050 as needed by the projected population increase. That would require reductions in canal and reservoir losses as well as significant increases in on-farm investments for better irrigation technologies and crop management practices. Rural-urban partnerships and national investments can certainly be a key part of those programs to balance these emerging and growing needs for increasingly scarce water resources.

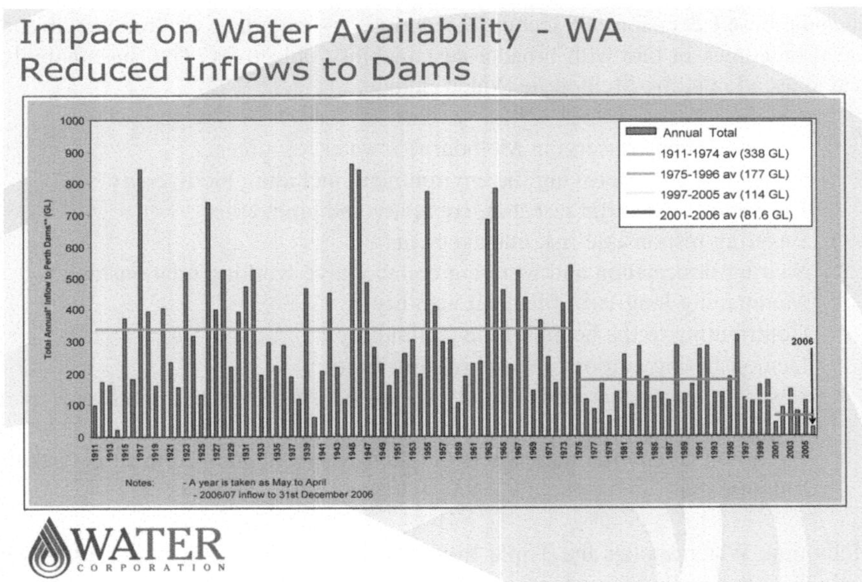

Figure 6.2 History of River Inflows to Dams. Source: Water Corporation (2009), Perth, Australia

6.3 SUSTAINABILITY - A PATH FORWARD

More governmental agencies and businesses are using sustainability principles that are not only socially responsible and protective of our environment but also promoting organizational advancement and sound fiscal stewardship. 'Sustainability' describes actions where the use of natural, community and utility resources are made in a manner that meets current needs without compromising the ability to meet future needs or approaches. Sustainable programs can also be described with objective-based goals that provide higher levels of reliability, resilience, efficiency and organizational adaptability and performance. These approaches should also facilitate better resolution of conflicting value systems in resource constrained settings when competing opinions or approaches are stalling or preventing the advancement of even superior and necessary technical proposals.

Tactical and strategic approaches to developing sustainable water systems are often characterized as meeting 'triple bottom line' objectives. To quote the comprehensive goals of the triple bottom line guidelines developed by Melbourne Water (2007): 'this constitutes a framework for decision-making that helps to provide guidance towards implementing activities that provide the greatest benefits to the community and the environment, while meeting the efficiency requirements for public spending'. This framework changes the focus from purely financial-based decisions of technical solutions to including environmental and social outcomes in line with broader sustainability objectives. The Sustainability Principles adopted by Melbourne Water include:

- Protecting and conserving Melbourne's water resources
- Protecting and improving the environment, including biodiversity
- Leadership, scientific research, creativity and innovation
- Ensuring responsible risk management
- Sharing information and fostering collaborative working relationships
- Maintaining long-term financial viability
- Contributing to the health of the community
- Demonstrating corporate social responsibility
- Ensuring inter-generational equity by considering short term and long term implications in all decision making
- Providing an environment where employees are encouraged to achieve their full potential

Melbourne Water applies the Triple Bottom Line concepts across the project life cycle from the preliminary stages of investigations and feasibility studies through detailed design, construction, operations and decommissioning.

Similarly, the Water Corporation in Perth, Western Australia (Water Corporation, 2009) applies eighteen Sustainability Business Principles to govern the conduct of its business. The principles are structured into six dimensions. The first three principles represent the traditional triple bottom line dimensions of social, environmental and economic outcomes – the 'Outcome Principles'. The principles acknowledge that current practices are not sufficient and that opportunities exist to improve the sustainable performance of the water system. These principles drive initiatives including programs such as water conservation, energy efficiency, renewable energy development, public education and awareness and indirect potable reclamation. The second three principles address ethical, stakeholder and governance issues and are considered as 'Process Principles'. These Process Principles recognize that higher degrees of sustainable performance can only be achieved if the manner in which issues are approached is itself considered thoughtful, transparent and collaborative. This second set of principles can expand beyond the technical approach to addressing water issues by describing project elements and programs that are more accepted by the community and policy makers. In the case of Water Corporation, this has led to the energizing of desalination plants using renewable wind generation and to the development of a major groundwater recharge program using highly treated wastewater. These two programs are elements of the more comprehensive *Water Forever* program that will provide sustainable water resources through mid-century, even with the move away from primary reliance on rainfall-dependent sources.

6.4 A FRAMEWORK FOR TRANSFORMATION TO SUSTAINABLE UTILITIES

More sustainable and effective utility operations can be identified by benchmarking approaches developed by water agencies in different regions of the world and that currently have to respond to the conditions that are expected to become commonplace in the decades to come. A recent review of Best Practices for Asset Management included structured interviews with 15 water agencies in the United States, 6 agencies in Australia and New Zealand and two private water companies (General Accounting Office, 2004). Asset management is considered a core business function of sustainable utilities and improves the sustainability of capital assets while enhancing decision-making related to project planning and collaborative relationships with communities, policy makers and stakeholders.

Benchmarking assessments of effective utility management practices (US Environmental Protection Agency, 2008) described ten attributes of water agencies that are seen to be essential to sustaining water infrastructure. Those ten attributes and their components are described in Table 6.1.

6.4.1 Key success factors

A series of key success factors influence management approaches and systems that are commonly used by water utilities guided by the attributes listed above. These include the following factors:

6.4.1.1 Leadership

Leadership and fostering a culture that supports continual improvement and effective change within the water provider's organization should be uniformly embraced by the policy or governance board, the executive management team and at all levels throughout the organization. Individuals and teams should ensure that their programs are broadly developed, information is effectively disseminated and supported and the programs drive operational and planning decisions

Table 6.1 Attributes of Effective Utility Management (after US Environmental Protection Agency, 2008)

Attribute	Components
Product Quality	Meets public health, customer and ecological needs. Complies with regulatory and reliability requirements.
Customer Satisfaction	Provides reliable, responsive and affordable services. Responsive to customer needs and emergencies.
Employee and Leadership Development	Collaborative organization led by integrated senior leadership with continuing recruitment and retention of competent and dedicated work force.
Operational Optimization	Minimizes resource use and losses through timely performance monitoring and improvements.
Financial Viability	Predictable and adequate pricing to cover life-cycle costs of debt, operations, rehabilitation and maintenance and capital improvements.
Infrastructure Stability	Maintain and enhance assets at lowest possible life cycle costs and with acceptable risks.
Operational Resilience	Effectively manages risk within established tolerances.
Community Sustainability	Efficient use of water and energy resources, promotes economic vitality, enhances natural environment and improves community sustainability.
Water Resource Adequacy	Manages long term reliable water supplies through public education, conservation and supply projects.
Stakeholder Understanding and Support	Active stakeholder involvement; Engenders understanding and support from policy makers, regulators, customers and watershed interests.

throughout the organization. The governance board and senior managers should be uniformly committed to serving as leaders in the industry in terms of their approaches to participative and effective community involvement and routinely search for better and more sustainable approaches. A measure of success in this area is when the community recognizes these inclusive approaches and can actively support proposed investments requiring tariff increases or in policy positions that may be unpopular but advisable to create a balance between competing interests. This desired relationship between the utility and the customer base is a core goal for Seattle Public Utilities (2004) as they developed their comprehensive asset management program: they are 'able to consistently meet the needs, desires and expectations of customers … through establishment and monitoring of service levels …. (and) application of triple bottom line decision making and … keep(ing) capital and operating costs low.'

6.4.1.2 Strategic Business Planning

An effective strategic business plan provides the context within which the water provider operates and identifies the mission, vision and goals of the organization. By identifying the risks and balances that are being achieved, customers and policy makers can help to focus on specific areas and degrees of investment, the expected levels of service and provide a documented and defensible basis for decisions that are made by the management team. Metrics and utility-specific benchmarking can help guide management decisions and provide the basis for changes in approaches as well as measuring and influencing improved performance levels. This will allow for continual improvements in key areas, provide the basis for decisions and levels of investment and provide the support for adaptive management and sustainability programs. This benchmarking can also develop a basis for describing the broader background and understanding of proposed programs that require significant political support. Major water utilities in Australia have noted that the use of sustainability principles and triple bottom line accounting have enhanced their relationships with regulators and stakeholders and provide an effective platform for developing consensus on prospective projects and operations.

The strategic business plan can also describe and promote the involvement and active participation of staff throughout the organization. Continual feedback from employees as well as various sectors of the community and interested stakeholders can be integrated into the strategic planning process to promote higher effective participation and a better understanding on how to meet future needs and to respond to identified deficiencies.

6.4.1.3 Effective Management and Governance

Strong and interactive leadership is required at multiple levels within the organization. It is also required externally within the regional or national framework as well as in the community that is served by the water provider. Leadership is exhibited by the development, coordination and functioning of a structured organization that has integrated responsibilities and accountability for reliably providing water services. Silo-ed or independent operations in different parts of the organization are typically not effective but are still commonplace. One of the consequences of the comprehensive asset management program developed by Seattle Public Utilities has been the integration of multiple functions in the utility: 'Establishment of service levels and performance indicators requires us to focus externally on customers and internally on our own productivity' (Seattle Public Utilities, 2004).

These utilities create a safe working environment for a motivated and empowered work force that creatively addresses the provider's operational and infrastructure needs and readily adapts to changing conditions. The water provider operates openly and collaboratively within their community and seeks regular feedback on measured performance so a continual improvement program can be used to enhance services over time.

The sustainable utility will consider a range of risks associated with climate change, source variability (drought, flood, diminished assimilative capacity in receiving environment), and disruptions (infrastructure failures, sea level rise) when assessing the natural resource needs and infrastructure framework required to provide the desired level of service. A resilient and reliable utility framework will also consider potential effects of regulatory, policy, security and environmental changes on their ability to serve their operations and projected needs. It will be increasingly important that water utility leadership be involved in the national and regional dialogue associated with these issues so they reflect the needs and reasonable actions by the water provider. Trade associations and professional societies are often used to collectively represent the perspectives of water agencies in legislative and regulatory matters.

6.4.1.4 Efficient Use of Resources

Comprehensive asset management programs such as those developed by Seattle Public Utilities (2004) have described the performance and condition of critical infrastructure. Those programs can be used to schedule and respond to potential and actual breakdowns or failures in ways that do not adversely affect service or that prevent such incidences. Infrastructure assets should be assessed and characterized on a regular basis and maintained, improved or expanded using full cost life-cycle approaches and by identifying acceptable limits of risk. Both tactical and strategic assets should be assessed within the context of current service levels and expectations as well as under projected growth and reliability/performance goals.

To manage utility operations and guide investment decisions, the sustainable utility will create a series of objective-based performance metrics that will reflect desired service levels, sustainable use of human resources, infrastructure and other collateral and the limitations under which the utility operates. Those benchmark standards could identify how resources such as consumption of water and energy can be minimized and would include various demand-side management programs and process adaptations. These efforts are often data intensive and so effective information technology and Infrastructure Management Systems that identify desired and time-sensitive actions are required. Sydney Water follows international best practice by mapping their activities against the Global Reporting Initiative (GRI) Sustainability Reporting Guidelines, (Sydney Water 2010). These reports are independently verified and provide managers with a baseline of performance that can be used to develop improvements programs.

Energy and water use by the water sector are closely linked. Integrating energy efficient practices into the daily management and long-term planning of the water sector can also contribute to long-term sustainability and managing the environmental footprint of water operations. Recent federal Executive Orders in the United States are requiring federal agencies to achieve significant energy and water efficiencies in the next several decades. Energy management programs in the Netherlands, Denmark and Finland are acknowledged by the United Nations as being some of the most successful examples that could be considered by others.

6.4.1.5 Full Cost Life Cycle Accounting

The magnitude of investments required to provide adequate water services indicates that sustainable utility operations should reconsider their traditional pricing, project delivery, service level reliability, capital and operational planning and regulatory/policy roles.

The level of investments projected to meet future needs cannot in most cases be supported under a traditional cost-of-service approach with the current pricing levels. Water and sanitation services are often undervalued and current pricing levels are often not adequate when the provider is limited to internally generating the revenues needed to support future operations and construction (US Environmental Protection Agency, 2002).

To be self-supporting, a utility would establish a rate or tariff structure that would serve as a full cost recovery enterprise and provide predictability and community support for the pricing structure that would cover costs as well as to plan and invest in future programs. Those rate requirements should be communicated to the customer base so their effects are predictable and supportable based on increased capacity and reliability to serve customers' water needs.

6.4.1.6 Integrated Resource Management

Sustainable utilities are actively involved in regional natural resource policy discussions and debates and help to form the framework in which water allocations and environmental regulations are established. For example, major water utilities are actively engaged with state and federal agencies in the identification of future management strategies on the Colorado River to respond to water shortages, international obligations and meeting the growth in demands within their water systems. Denver Water, Colorado has been engaged in ongoing negotiations with distant basins of origin to resolve differences in the legal availability of future water supplies and in identifying acceptable mitigation measures that would allow the development of those water resources to occur.

Increasingly, sustainable utilities are creating an open and participative dialogue with the public and non-governmental organizations on their programs. They are soliciting input on acceptable approaches as they consider future initiatives including source development, rate structures and desired levels of service. Comprehensive demand-side programs are now seen as a first step of integrated planning. Demand-side management includes comprehensive water conservation programs, loss and leak management, support of land use and community planning approaches, expansion of water reclamation programs and public education and awareness.

Integrated resource management approaches provide an assessment of what could be supported by the community and policy makers. Wide-ranging community involvement programs through to an effective utility governance approach could be needed to understand the local circumstances as well as national and regional agendas.

In a recent benchmarking assessment (Binney and Lampe, 2009) a number of Australian, English and American water utilities described how they have responded to their service needs with customized integrated water resource management programs. The Southern Nevada Water Authority serves the rapidly growing Las Vegas metropolitan area. Their program includes one of the most aggressive water conservation approaches in the United States with an emphasis on reducing outdoor water consumption which had represented about 60% of the annual demand on the water system. Large tracts of irrigation-dependent landscapes are being replaced with low water-demanding xeriscape areas. These programs have resulted in lower water demands in 2007 than in 2002 even though an additional 400,000 new residents were added to the service area. Because of limited availability of water from the Colorado River, the water agency has planned or developed an interstate groundwater banking program with Arizona, augmentation of river flows by importing remote groundwater resources and developing substitute supplies by exchanging new downstream sources up to their diversions at Lake Mead. Lower reservoir pool elevations caused by protracted drought conditions have also required the addition of deeper intakes to allow continued pumping to the water system.

Florida's Tampa Bay region has responded to drought and water scarcity through a number of approaches. While primarily dependent on local groundwater resources through the 1990's, the water utility has moved strategically to supplement reduced pumping with the development of surface water resources to limit the environmental impacts of lowered groundwater levels. This has included an updated conservation and demand management program, state of the art water treatment plant, a large surface water storage reservoir and the largest seawater desalination plant in the United States. The utility is also maximizing the use of reclaimed water to reduce the demand for higher quality source waters that should be preferentially used for human consumption.

The Water Corporation serves the water needs of one third of the area of Australia. Outside of the 1.6 million people in Perth, most of the state is sparsely populated. Despite ongoing severe drought conditions around Perth, water use has been reduced by improved irrigation practices and increases in urban supplies from a seawater desalination plant. Water efficiency initiatives such as education programs, building code changes, more efficient plumbing and irrigation systems and increased-density living are expected to reduce demand. New sources of surface water supply, increased operational efficiency of water stored by existing dams, sustainable groundwater sources, cloud seeding, additional recycling, catchment management, water trading and future desalination of sea water are also included in the 50-year *Water Forever* plan. The utility is also assessing the feasibility of replenishing groundwater aquifers using highly treated wastewater.

Aurora, Colorado faced a major reduction in service levels because of severe drought conditions in 2002 and 2003. Their comprehensive response included a water conservation program that reduced annual water demands by more than 35% and upgrades to treatment plant processes to allow increased capacity to deliver marginal quality source waters to the system. The city also developed a continued farming program where irrigators could periodically fallow lands and lease irrigation water for municipal use until more normal water supply conditions returned. The cornerstone of the city's response was a $750 million program to recover highly treated wastewater from downstream locations through a river bank infiltration system and then pumping the recovered water back to an advanced water purification plant before delivery to the drinking water system. The purification plant includes softening, advanced oxidation, ultra-violet, filtration and disinfection processes to maintain the high quality of water that customers have received for decades from snowmelt.

These various case studies point to the comprehensive approaches that are being taken by leading edge water utilities as they successfully respond to the challenges described earlier in this chapter. This developing body of knowledge can be considered by other water agencies as they develop their unique approaches to sustainably meeting their own water needs.

6.5 CONCLUSIONS

Increasing complexity and dynamic changes facing water providers could easily overwhelm traditional approaches to utility operations and capital investment programs in the water sector. A benchmarking review of a number of providers in different regions of the world indicate a series of key success factors that are being used to create sustainable utilities that will remain relevant and effective in the coming decades. Other water providers can consider their own situations and the approaches that are being used by those sustainable utilities to establish their own adaptations to future operations and capital investment planning.

REFERENCES

Binney, P. and Lampe, L. (2009). 'Water Utilities Cope With Water Scarcity Issues', World Water and Environmental Engineering (January 2009) pp. 15–17, Volume 32, Issue 1.

Booz, Allen, Hamilton (July 2007). The Megacommunity Way: Mastering Dynamic Challenges with Cross-Boundary Leadership.

General Accounting Office (March 2004). Water Infrastructure – Comprehensive Asset Management Has Potential to Help Utilities Better Identify Needs and Plan Future Investments– GAO-04-0461. Washington D.C., USA.

Melbourne Water (2007). Triple Bottom Line Guidelines, Melbourne Water, Melbourne, Australia.

Natural Resources Defense Council (2010). Evaluating Sustainability of Projected Water Demands under Future Climate Change Scenarios.

Seattle Public Utilities (2004). Asset Management - Seattle Case Study. Seattle, Washington, USA.

Sydney Water (2010). Sydney Water Annual Report 2010. http://www.sydneywater.com.au/annualreport/sustainability/global_reporting_initiative.html Accessed 12 March 2011.

United Nations (2000). UN Millennium Declaration, UN A/Res/55/2, 2000.

US Environmental Protection Agency (2002). The Clean Water and Drinking Water Infrastructure Gap Analysis (Executive Summary and Conclusion), EPA-816-R-02-020 (September 2002), US Environmental Protection Agency, Washington D.C., USA.

US Environmental Protection Agency (2008). Effective Utility Management – A Primer for Water and Wastewater Utilities. Washington D.C., USA.

US Environmental Protection Agency (2009). Drinking Water Infrastructure Needs Survey and Assessment, Fourth Report to Congress, EPA 816-R-09-001, (March 2009), US Environmental Protection Agency, Washington D.C., USA.

US Geological Survey (2009). Estimated Water Use in the United States – 2005, USGS Circular 1344. Washington D.C., USA.

Water Corporation (2009). Water Forever, Perth, Australia.

Part II

Bringing the People with You

Chapter 7

Communicating across disciplinary divides – are we bridging the gap?

Carol Howe

UNESCO-IHE Institute for Water Education, Delft, The Netherlands

ABSTRACT

Sustainability is a concept that requires consideration of economic, environmental and social benefits and impacts. Most people find it hard to grasp one of these elements let alone understand the balance and interactions between all three. Interdisciplinary and transdisciplinary research as well as integration across the water cycle are concepts that are achieving more traction as means to ensure sustainable outcomes.

As we attempt to move towards this more sustainable form of decision-making there are questions around whether specialists, especially academics, have the nuanced and subtle communication skills and generous mental attitudes required to work together in an open, equal and respectful way. If we do not, how can we expect our teams who are often developing complex and innovative concepts and solutions to be able to communicate these ideas externally? How can we expect stakeholders to trust or understand our decision-making processes?

This paper explores how engineers, environmental scientists, social scientists and economists process information and communicate with others. Reasons for differences between disciplines including academic training (epistemologies), personality types and motivations will be discussed. The author draws upon her experiences as project manager of the European Union's SWITCH (Sustainable Water Improves Tomorrow's Cities Health) research project to illustrate how different disciplinary ways of thinking and communicating are affecting the working environment, direction and outcomes of a large (over 150 researchers in 17 countries) and theoretically integrated urban water research programme.

Finally, suggestions are given on how multidisciplinary groups working on urban water management might transition to truly integrated and transdisciplinary teams.

Keywords: communication; decision making; interdisciplinary; learning alliances; transdisciplinary

7.1 INTRODUCTION: WICKED PROBLEMS AND COMPLEX SYSTEMS

Many cities throughout the world are struggling to manage water effectively and many more will struggle in the future due to increasing pressures from climate change, population growth and migration, energy shortages and fluctuating economic conditions. Water management in both developing and developed countries often focuses on today's problems using conventional approaches. Such approaches are often fragmented, with components of the various elements of the urban water system designed and operated in isolation. The science that underlies decision making also tends to be carried out in isolation.

An increasingly common requirement of agencies funding water management innovation is for researchers to ensure that their work is driven by demand and effectively communicated. The underlying rationale is to improve the impact of research on policy and outcomes. As a result, individuals and projects are under pressure to do much more than what was traditionally understood as 'good fundamental or core science'. They are required to not only understand the priorities of potential users and their institutional context, but also to undertake research collaboratively across scientific domains and to develop and communicate new, integrated and innovative solutions.

The concepts underlying complexity science provide researchers and practitioners with some of the tools that are needed to come to terms with the dynamics and change processes of complex systems. Complexity science initially focused on physical and biological phenomena, but it has gradually started to gain ground in the social, economic and political sciences. 'Wicked problems' (the term coined by Rittel & Webber in 1973), common in complex systems, are defined as problems that can only be understood by collaboratively exploring solutions. Each wicked problem is new and unique, there are no definite solutions and solutions are not right or wrong. Conklin (2006) explains how wicked problems in complex systems often lead to fragmentation. Fragmentation suggests that people consider themselves as separate rather than united in a situation in which knowledge and information is scattered. The fragmentation essentially represents the different views from different stakeholders that all feel their view is the most correct and their problems are the most urgent.

In recognition of these linkages between society and the environment, projects more frequently use terminology such as sustainable, integrated (water resource management, coastal management, etc.) and interdisciplinary to describe their research. The EU project AquaStress Integrated Programme states it will 'deliver *enhanced interdisciplinary methodologies* enabling actors at different levels of involvement and at different stages of the planning process to mitigate water stress problems.' CSIRO (Australia) Water for a Healthy County states that it 'provides science and technology that enables the *delivery of socially acceptable, affordable and environmentally beneficial* water management solutions for Australia's cities'.

Are these statements just rhetoric or are researchers actually working together to solve problems? This article explores some of the realities of interdisciplinary or transdisciplinary research and provides some recommendations on how systematic solutions that bridge the economic, environmental and social aspects of water management might be better realised. The five-year EU research project SWITCH is used as a case study. SWITCH aims to achieve more sustainable urban water management in the 'city of the future' by having 33 organisations work together on innovative scientific, technological and socio-economic solutions.

7.2 UNDERSTANDING THE TERMINOLOGY

> *'The most exciting science in the 21st century is likely to evolve among not within traditional disciplines ... yet the education of scientists has historically been constrained by disciplines, paralleling patterns of science funding.'* (Sung *et al.*, 2003)

Often the terms multidisciplinary, interdisciplinary and transdisciplinary are used interchangeably, despite having different interpretations. An understanding is however emerging that, as we move up the scale from multidisciplinary to transdisciplinary, the interaction and connections between disciplines and stakeholders increases. Tress *et al.* (2004) define projects as interdisciplinary if they involve several unrelated academic disciplines in a way that forces them to cross subject boundaries to create new knowledge and theory and solve a common research goal. By unrelated, they mean that they have contrasting research paradigms, such as differences between qualitative and quantitative approaches or between analytical and interpretative approaches, that bring together disciplines from the humanities and the natural sciences. They define transdisciplinary studies as 'projects that both integrate academic researchers from different unrelated disciplines and non-academic participants, such as land managers and the public, to research a common goal and create new knowledge and theory. Transdisciplinarity combines interdisciplinarity with a participatory approach.'

In increasing order of level of interactive problem-solving, we define the following terms:

- disciplinary: specialist in isolation
- multidisciplinary: investigate single problem but from own discipline; stapling together of results
- interdisciplinary: coordination from higher level concept, unified problem formation, sharing of methods
- transdisciplinary: transcends entrenched categories to formulate problems in new ways; coordination across all levels and usually with engagement of society.

The core idea of transdisciplinarity is that different academic disciplines work jointly with practitioners to solve a real-world problem (Klein *et al.*, 2001). There is also a third group: the community or broader stakeholders that some consider an important component of transdisciplinary work (Wickson *et al.*, 2006).

7.3 DISCIPLINARY APPROACH TO SUSTAINABILITY

'The world has problems, but universities have departments.' (Brewer 1999)

Although joint problem-solving is a high ideal, in reality the focus of problem-solving is generally weighted towards a specific discipline and therefore a specific aspect of sustainability. Most interdisciplinary research ends up entitling a single discipline, incorporating others in a support or service role; this is referred to as 'epistemological sovereignty' (Healy 2003). In urban water management, this has historically been dominated by engineers.

Different epistemological and methodological paradigms have competing views on why a problem exists in the first place, how it has manifested and what the best ways to arrest or address it are. A profound challenge is to work across epistemological divides in order to meet the need to articulate between different forms of knowledge (Thompson-Klein 2004).

Figure 7.1 illustrates what aspect of sustainability different disciplines typically focus on, in their approach to a problem.

Engineers and environmental scientists take a positivist approach of systemised knowledge relating to criteria of what ought to be. They rely on quantitative data and models for problem solving.

According to Roughly & Salt (2005), 'Social sciences frame the context in which other knowledge can be applied; questioning the fit between that knowledge and its context and evaluating its purpose; and providing a critique of science and technology which is valuable as an input to technological decision-making from the beginning'. In Australia, social scientists with the National Resource Agency introduced qualitative and participatory approaches to organisations with a natural science understanding. These research and management models were often viewed in the agencies as not scientific, being less rigorous than positivist science. Qualitative, participatory approaches were considered to be not a matter of science at all but of common sense: 'When it comes to the social, everyone's an expert'. Three managers identified this as one of the most challenging barriers to effectively integrating social science within natural resource management (Roughley & Salt 2005).

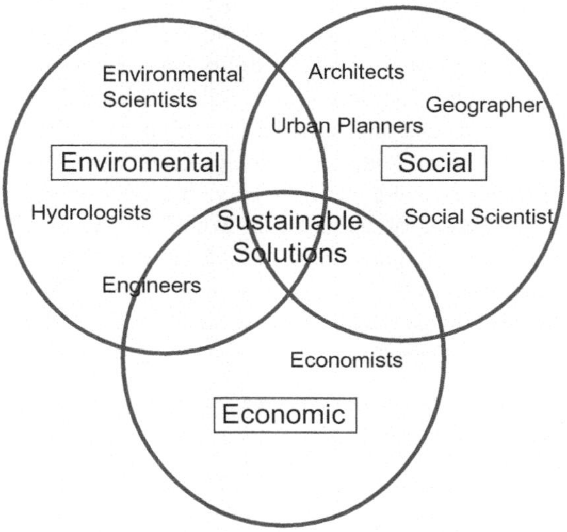

Figure 7.1 Disciplinary focus for sustainability

There is often an unrealistic expectation regarding social science. Expectations range from getting results quickly in changing community attitudes, to developing a system for engaging the community. Neither of these goals is quickly attainable. Expectations are generally quite vague, but usually include:

- engage, educate and at times placate the community
- directly undertake, design or manage research
- play a key (if not lead) role in establishing multi-disciplinary integration across the agency.

Social scientists are often not valued for their actual science, but viewed as general facilitators for other disciplines. As G. Syme (a respected sociologist at CSIRO) said in the Roughley & Salt (2005) report: 'Research should assist in addressing the community's problems not just supply information about the community for the purposes of the research of biophysical scientists. It should be a partnership, not paternalism.'

Economists span the area between social and physical sciences, taking both qualitative and quantitative methods into account. They could provide a natural bridge between sciences but often reduce the world to cost–benefit equations that, from the author's experience, are often distrusted by both hard and soft scientists alike.

A city that is going through a process of redesigning or developing its urban water system for the future would obviously like to know where it is going. Is it getting nearer to sustainability or not? One has to recognise that sustainability cannot be measured in an absolute way, such as the pH of a water sample. Sustainability is in fact defined by the stakeholders who give different weights to different aspects, and perception of sustainability can change over time.

Indicators can be used to measure sustainability: sustainability indicators (SIs) have been designed from existing knowledge within each discipline of science, being the origin of diverse approaches. Each approach seeks to understand what a sustainable society means, using SIs as a way to assess whether a development pattern leads to sustainability as a goal. Some of these will be based on qualitative data and some on quantitative and, as shown in Figure 7.2, include the use of a variety of assessment tools such as Life Cycle Assessment, embodied energy, behaviour/attitude investigations and cost–benefit analysis.

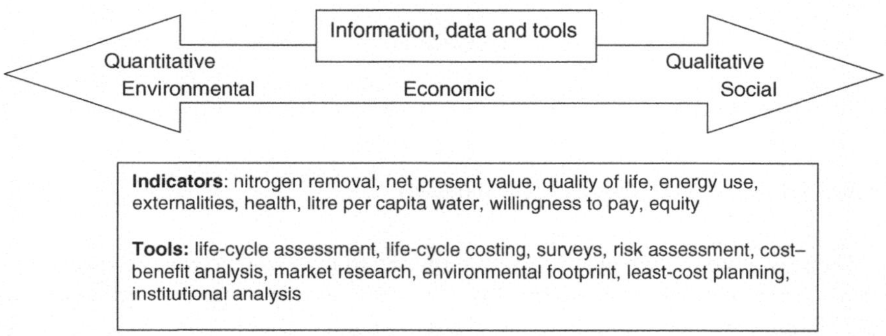

Figure 7.2 Example indicators and assessment tools used by different disciplines

A researcher must be open and equitable in the use of a wide range of tools and results to get as complete a picture as possible of the sustainability of the system. Various types of overarching multicriteria decision-support tools are now emerging, but the value of these is only as good as the balance and depth of information that is input to the analysis.

7.4 SWITCH: A PROJECT DESIGNED FOR INTEGRATION AND SUSTAINABILITY

In SWITCH a conscious effort was made to recruit researchers and practitioners from cities around the globe that covered a diversity of climate, socio-economic and institutional situations. The project was focused on improving the sustainability of urban water management in these cities and, by having this diversity, it was assumed that replication of knowledge would be accelerated. However, when it came to the make-up of the consortium, selection was based more on the 'hard' science areas of water management with an almost 4 to 1 ratio of 'hard' scientists (engineers, scientists, etc.) to 'soft' scientists (sociologists, economists, urban planners, etc.) (Figure 7.3). This initially caused considerable tension in the project, with both groups frustrated with each other. The social scientists wanted to invest time and funding into the process of forming "learning alliances" in the cities including understanding of the stakeholders, power relationships and institutional structures. The engineers and physical scientists wanted to get on with their research and were reluctant to allocate additional funding to engagement processes.

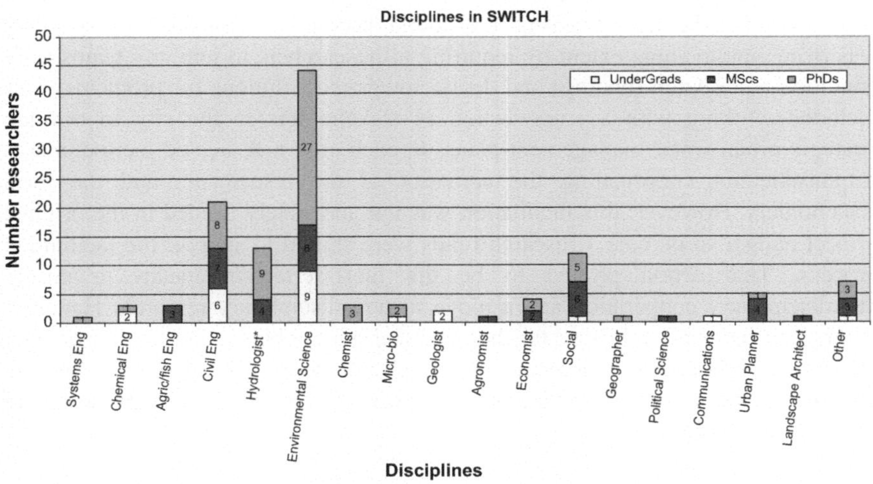

Figure 7.3 Distribution of disciplines in the EU SWITCH project

Even within disciplines, particularly in the physical modelling and economic areas, there were differences of opinion on methods and approaches. Communication between social scientists and physical scientists was particularly interesting to observe. Often the social scientists would write detailed and persuasive emails only to have a one-line response, basically asking 'and your point was?' The response of one of the project's engineering theme leaders to a social analysis study of water-sensitive urban design was typical of the engineers' responses: 'thanks, interesting...although the outcomes are rather general/vague... I would like a more structured process'.

Working in interdisciplinary teams requires openness and patience. The language of the different disciplines varies markedly, with each discipline having its own acronyms and terminology. For example, environmental scientists might talk about life-cycle assessments and social scientists about capacity building. It is easy to tune-out in a discussion when you do not understand the basic language. People involved in interdisciplinary work must be particularly attentive to the different claims, theories and methods of different research domains. To move a project from multidisciplinary (no cooperation) to interdisciplinary (coordination from higher level concept) to transdisciplinary (coordination across all levels), and consequently to one of true integration, requires working across epistemological divides to articulate different forms of knowledge and for researchers and practitioners to take on new roles to facilitate the process (Figure 7.4). Mitchell (2007) states that 'transdisciplinary research is personally challenging, requires humility and seeks to integrate different ways of knowing from a non-denominational starting point'.

In SWITCH the natural resistance of researchers to work outside their specialty was overcome to some extent by requiring all researchers to join city teams (as a condition of research funding) and devise integrated solutions for problems. This included working with a cross-section of stakeholders in the city to develop strategic urban water management plans. In each city, a dedicated facilitator was responsible for coordinating the activities of the researchers and the local stakeholders. However, this facilitation was not adequately funded in the original project budget so in Year 1 research funds were shifted to support the facilitation process. This action proved to be the most effective means to ensure interdisciplinary cooperation and a move to transdisciplinary research. Having a strong mechanism to bring stakeholders and researchers together in the city through well-thought-out 'learning alliances' and tying this to research funding proved to be a strong motivator for working together on exploration of new solutions (Figure 7.4).

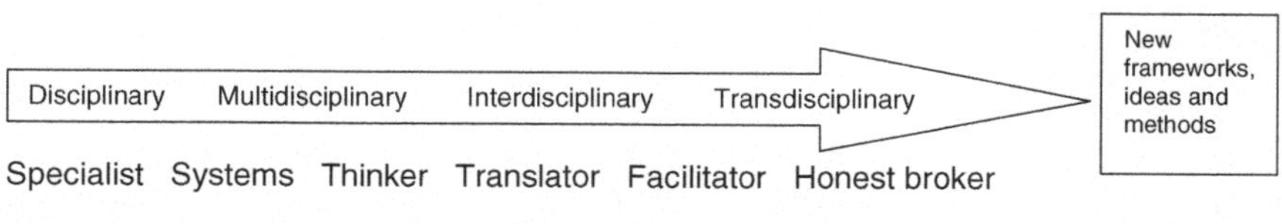

Figure 7.4 Skills sets needed to move towards transdisciplinary action

Each city had a different problem, for example water scarcity, flooding, lack of sanitation or other issues. In SWITCH a core group of researchers tackled this immediate problem with local stakeholders while working with a wider group to integrate this problem and its solutions into a wider water management plan.

7.5 RECOMMENDATIONS

7.5.1 Changes to the education system

Mitchell *et al.* (2004) contend that traditional teaching, learning and practice in engineering and applied science has largely been confined within the techno-centric sphere, having minimal interaction with the eco-centric and socio-centric concerns.

Meeting the challenge of tackling global problems requires today's students to be equipped with both traditional core skills and new skills such as facilitation and insights from a broad range of disciplinary areas including the science of complexity, sociology, philosophy, law, ecological economics and management. Robinson (2008) suggests that there is much resistance to issues-based interdisciplinarity in the academic world, both principled (standards, quality, etc.) and rooted in practice, in the form of the typical subject-matter, employment and reward systems. There are also concerns about academic freedom, lack of publishable work and the time-consuming and risky nature of the work. If sustainability is to be achieved and universities are to address large societal problems, then interdisciplinary work needs to be fostered. According to Scott (2002), learning how to learn is the single-most important goal for sustainable development. He proposes principles for learning how to learn (versus what is taught), including helping the learner understand why consideration of sustainability is in their interest and helping learners to gain a professional awareness of context (both their and others') as well as personal awareness (a view to our and others' values).

7.5.2 Better balanced teams

Often proposals for research and work are prepared by groups of people who have worked well together previously; we are naturally drawn to people who think like we do. In the case of urban water this is usually engineers, hydrologists and environmental scientists. To move towards new solutions requires a conscious effort to recruit people outside of our usual circle (social scientists, economists, urban planners, geographers, etc.) to achieve the necessary mixture of skills and creativity to break through traditional thinking. To be effective requires true balance versus a token economist or social scientist.

7.5.3 Training of teams

The early training of teams in the different disciplinary ways of thinking and processing information can help to foster interaction and understanding. If this can be done in a humorous and fun way, so much the better. Although there are many experts in the world, there are not many experts who will take the time to interact with other researchers outside their core discipline. Methods vary greatly between disciplines, and trust can quickly erode and undermine a team's efforts if disciplinary methods are not appreciated.

Techniques such as problem-based learning (PBL), where the process of addressing the problem is at least as important as the outcome, should be encouraged. In this approach the problem may be 'solved' at the point that the relative merits of a range of preferred options are communicated. PBL can also be useful when multidisciplinary teams are brought together and techniques such as role playing are employed. This may be challenging for teachers who are accustomed to being provided with one right answer. The point is to find the right question and to develop the capacity to answer those questions in an insightful, critical and consultative way. PBL naturally lends itself to the development of honest brokers (Mitchell *et al.*, 2004). To be able to tackle global problems, universities will need to place a greater emphasis on social learning processes, capacity building and interaction with other disciplines.

7.5.4 Appoint a neutral lead facilitator/coordinator/translator

Appointing someone to act as a neutral lead facilitator, coordinator or translator can bring people together and facilitate interactions between disciplines – invaluable to more sustainable outcomes. Without this facilitation, the process depends on the hope that interactions will occur; however, experience has shown that this is unlikely to be the case. Each researcher becomes absorbed in their own specific piece of work and, without a catalyst, interaction is unlikely to occur in any sort of meaningful way. Translation of terminology and methods is also critical; I have often sat through discussions where economists or social scientists explain their position or approach to a problem only to have my mind drift off because what they are saying is not in terms that are meaningful to me.

7.5.5 New learning environments

It is important to establish new learning environments, where we move away from defending perspectives and move towards shared insights and more 'systems thinkers' (systems engineers, ecological economists, etc.) who can help interpret and connect different languages and methods. Finally, to be truly successful requires common respect and appreciation of each other's values, methods and perspectives. This does not happen quickly, but requires patience and an open mind.

7.6 CONCLUSIONS

Robinson (2008) identifies issues-driven interdisciplinary research as having a very strong focus on partnerships with the external world. These partnerships go beyond treating partners primarily as audiences, and instead involve them as co-producers of new hybrid forms of knowledge. Over the past several decades, issue-driven interdisciplinary work has become especially prevalent in the field of sustainability. There is no one dominant discipline in this realm, thereby providing a real opportunity for the water industry to provide leadership.

Engineers, scientists and other specialists are not traditionally known for their strong communication skills. Each individual will have their own comfort zone for external communication across the realm, ranging from 'honest broker' within their own spectrum of expertise to full facilitator of transdisciplinary processes. It is not necessary for everyone to be an accomplished communicator, but it is necessary that researchers talk with a coordinated voice about integration and sustainability.

When research is communicated well, positive experiences result. Integrated inter- or transdisciplinary approaches reveal that integration of knowledge occurs most effectively in small, non-hierarchical structures where team members have equal status and the leader is a generalist in orientation (Rossini & Porter 1984; Klein 1990; Booth *et al.*, 2001).

Integration across disciplines occurs when problems are formulated in terms that enable the different disciplines to work in close collaboration without competition, and is facilitated where the method of data acquisition is helpful to all the relevant disciplines (not biased in a particular direction).

If we are going to move towards sustainability in urban water, then we all need to be reflective of the way that we view and analyse the world and realise the limitations of our own knowledge and experiences. Any hope of sustainability will require a multitude of views and approaches to ensure that we are on the best path as well as an acceptance of the necessity to find accommodation among the differences.

REFERENCES

Booth, M., Frost, R., Rodgers, S. & Bukerup, A. 2001 Interdisciplinary Research Methodologies in Natural Resource Management, Land & Water Australia Project Research Paper No. UMU14.

Brewer, G.D. 1999 The challenges of interdisciplinary. *Political Science* **32**, 327–337.

Conklin, J. 2006 Dialogue mapping: Building shared understanding of wicked problems. Wiley. Available at http://cognexus.org/wpf/wickedproblems.pdf.

Healy, S. 2003 Epistemological pluralism and the politics of choice. *Futures* **35**(7): 689–701.

Klein, J.T. 1990 Interdisciplinarity, History, Theory, and Practice. Wayne State University Press, Detroit.

Klein, J.T., Grossenbacher-Mansuy, W., Häberli, R., Bill, A., Scholz, R.W., & Welti, M. (Eds.). (2001). *Transdisciplinarity: Joint Problem-solving among Science, Technology, and Society*. Basel: Birkhäuser.

Mitchell, C. 2007 Zen and the Art of Transdisciplinary Research: An Overview, Public lecture, Chalmers University of Technology, 8 May 2007, Fakultetsvåningen, Villan Gibraltargatan 1A. Chalmers University Public Lecture.

Mitchell, C.A., Carew, A.L. & Clift, R. 2005 The role of the professional engineer and scientist in sustainable development in practice: case studies for engineers and scientists. In: A. Azapagic, S. Perdan, & R. Clift (Eds.), *Sustainable Development and Practice*. John Wiley & Sons, Ltd, Chichester, UK. doi: 10.1002/0470014202.ch2.

Porter, A.L. & Rossini, F.A. 1984. Interdisciplinary research redefined: multi-skill, problem-focussed research in the STRAP framework. *R&D Management* **14**, 105–111.

Rittel, H.W.J. & Webber, M.M. 1973 Dilemmas in a general theory of planning. *Policy Sciences*, **4**, 155–169.

Robinson, J. 2008 Being undisciplined: transgression and intersection in academia and beyond. *Futures* **40**, 70–86.

Roughley, A. & Salt, D. 2005 Introduction of social sciences in Australian natural resource management agencies. *Journal of Research Practice*, **1**(2), Article M2.

Scott, W. 2002 Education and sustainable development: challenges, responsibilities, and frames of mind. *The Trumpeter* **18**(1), ISBN 0832-6193.

Sung, N.S., Gordon, J.I., Getzoff, E.D., Kron, S.J., Mumford, D., Onuchic, J.N., Scherer, N.F., Sumners, D.L., & Kopell, N.J., 2003 Educating future scientists. *Science* 12 September 2003: 1485. [DOI:10.1126/science.1086133].

Tress, B., Tress, G. & Fry, G. 2004 Defining concepts and the process of knowledge production in integrative research. In: Tress, B., Tress, G., Fry, G., & Opdam, P. (Eds.), *From Landscape Research to Landscape Planning: Aspects of Integration, Education and Application, Proceedings of the Frontis Workshop*. Springer, The Netherlands, pp. 13–26.

Thompson-Klein, J. 2004 Prospects for transdisciplinarity *Futures* **36**: 515–526.

Wickson, J., Carew, A.L. & Russell, A.W. 2006 Transdisciplinary research: characteristics, quandaries and quality. *Futures*, **38**, 1046–1059.

Chapter 8

Plain speaking about water – experience from the trenches

Jennifer Simpson

Queensland, Australia

ABSTRACT

When we consider how to manage water so that it is sustainable into the future in the face of climate change and increasing population, it is vital that the full palette of all possible water and wastewater management options is identified and evaluated. At the moment the advantageous option of recycling to augment our drinking water supplies is not being considered as many proposed projects have failed due to public opposition. Other solutions that are more expensive and energy consuming are being implemented instead. Negative community attitudes to recycling are heightened by the language of water. Much of it is jargon, technical terms and acronyms; many of the terms are stigmatising and create alarm. It is not suitable for explaining wastewater and its treatment to the layperson. There is a vacuum in community knowledge about water that leaves water management issues open to political exploitation. The products from a wastewater treatment plant are described in terms of their degree of treatment – their history – whereas the community wants to know what they can safely be used for. There are also anomalies – why is recycled water alright if it is 'inadvertent' or 'unplanned' but not if it is done deliberately using modern best practice technology? In Australia the community is given the impression that planned recycling is more risky than unplanned recycling because it has separate and more rigorous guidelines. The insistence on an environmental barrier makes the community think there is something wrong with the water, as it needs to be 'diluted'. Information about recycling is given after the decision to recycle has been made but basic knowledge of water should be provided in advance – you cannot retrofit an education program. The community needs to know what is in wastewater, how it is taken out again and how they can be sure it has been taken out but recycling information invariably starts with the advanced treatments – the basics are omitted. Lack of community knowledge is not being addressed – such attempts that have been made to inform the community have been ineffective. A proactive water science education program is needed that simplifies the concepts and uses easy to understand words and images. This article points out the issues that need to be resolved if the community is to accept potable water recycling.

Keywords: Public opposition; negative terminology; misleading messages; community wastewater education

8.1 INTRODUCTION

I first became aware of the water industry in 1992, when there was a proposal to build a water supply dam in the valley where I lived. I thought it was outrageous to do such irreversible environmental damage to a lovely valley; surely there must be some other way of providing the water for our increasing population. I remembered being told that water in the River Thames went through seven people on its way to becoming the drinking supply for London. If they could successfully recycle water in the UK, surely we could do the same in Australia.

I needed to know about wastewater but I had the greatest difficulty finding information. There was a brief reference to wastewater treatment in the Encyclopaedia Britannica alerting me to the fact that it was a biochemical process. There was nothing in the library. It was, of course, before the days of websites. The only way was to visit wastewater treatment plants. Since those early days I have inspected over 100 of them, in 10 different countries.

The language of water is mainly jargon, technical terms and acronyms that have evolved over the years. In the past there had been no need or opportunity for water and wastewater engineers to share their knowledge with the community. Because of this the language of water is not suitable for explaining water quality and treatment to a lay audience. Many of the terms are stigmatizing – the language of water and regulatory processes create alarm – they actually scare people away from reuse! I learned that lack of knowledge of water science is widespread and extends to some senior members of Government Departments and decision-makers. Much of what they think they know is negative and they are often wrong. It is difficult to consult with the community about water issues as uninformed people have uninformed opinions.

This lack of knowledge leaves water management issues vulnerable to political exploitation. Political opportunists fill the vacuum in the community's knowledge with emotive statements that foster disgust. For example, Snow Manners and John Dowson of Toowoomba published a booklet entitled '*Think before you agree to drink – is sewage a source of drinking water?*' (Figure 8.1) In it they describe wastewater: '*We should remind ourselves about what goes into a sewer. Domestic, commercial and industrial toxic chemicals together with both prescription and illicit drugs from human waste or from dumping unwanted or out of date drugs down toilets, tubs and kitchen sinks. This joins liquid waste from morgues, hospitals and any other place connected to a sewer as well as paints, solvents and acids. Sewers are used to dispose of all manner of substances which arrive at the city's sewage treatment plant as one massive, horrendous toxic cocktail. Are the risks of using recycled sewage water for drinking really worth the possible consequences that may not be known for decades?'*

Figure 8.1 Cover of Think Before You Drink" Booklet

Water quality issues have the potential to form strong, negative headline material, which has great attraction to the media – it is easy for them to generate hysterical headlines that sell newspapers.

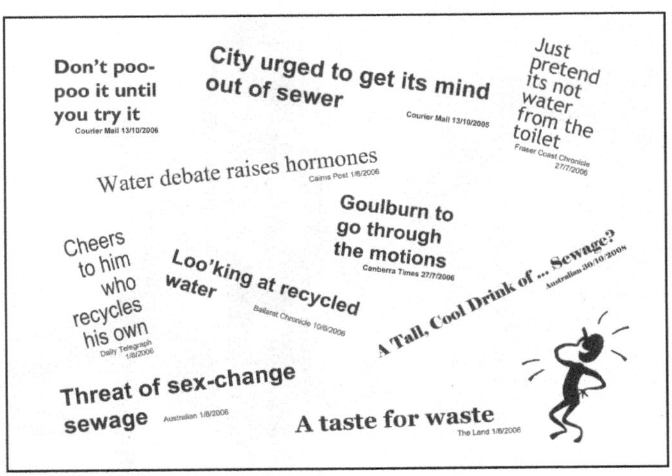

8.2 NEGATIVE TERMINOLOGY AND IMAGES

8.2.1 Would you like a glass of treated sewage, dear?

It is not surprising that talk about recycling sewage, wastewater, treated wastewater, treated effluent or treated sewage evokes feelings of disgust and receives a negative response. I have even heard of recycling treated sewerage. In New South Wales they are undertaking desalination rather than recycling because they say that people won't accept recycling. This was proven by a survey in which the question was '*would you drink recycled sewage that includes water from the toilet?*' (Sydney Water Survey, 2006).

The term '*recycled wastewater*' is frequently heard. At the AWA Water Recycling Conference in Sydney 2007 it was in the title of the paper by the Managing Director of Sydney Water and in a report on recycling by the National Water Commission. The conference, when reported in AWA's magazine, *Water*, was flagged on the front cover as Wastewater Recycling. The European project, Aquarec, was entitled, *Integrated Concepts for Reuse of Upgraded Wastewater.* When the Queensland Department of Natural Resources undertook its Water Recycling Strategy, it was originally called the Wastewater Reuse Strategy; it took 9 months of persistent nagging and, eventually, a decision by the Minister to change it.

The question '*would you like a glass of treated sewage, dear?*' was asked by a social researcher on the ABC National TV Four Corners program. The negative response was said to be typical of the 'yuck factor'. We hear so much about the yuck factor that we feel as if we are expected to be affected by it – we are not 'normal' if we are not. It is another example of terminology stigmatizing recycling. The 'yuck factor' is fully exploited by the media.

These expressions contrast with the terminology used by purveyors of bottled water to sell their product – clear, sparkling, fresh water from mountain springs.

8.2.2 Quality, not history

Dr Lucas van Vuuren (National Institute of Water Research, South Africa) a pioneer of water reclamation during the 1970's, said *'Water should not be judged by its history but by its quality'*. Although this is generally applauded, water professionals continue to place the emphasis on history, describing water quality in terms of its source and the degree of treatment it has had. This does not inform the community what the water can safely be used for. The terms, primary, secondary, advanced secondary and tertiary treatment are not well defined and have no meaning to the layperson. From the Western Corridor Recycling Project's glossary we found (since removed) an enlightening definition of secondary treated water: *wastewater that has been subjected to primary and secondary treatment at a wastewater treatment plant.* (www.gwc.gov.au).

The community is not interested in where our water has come from or how much treatment it has had. They would like to know, please, what it can be safely used for.

8.2.3 Recycled water: do not drink

On Queensland's Gold Coast residents are told that the water supplied in purple pipes in the Coomera dual reticulation scheme is *'Recycled water: Do not drink'*. They have the perception that they should not drink the water because it is **recycled**, not because it hasn't had the next stage of treatment that would make it safe to drink. Next door, in Brisbane, people are told *'Purified recycled water: Do drink*[1]'. Adding the word 'purified' has not overcome the problem of the perception of recycled water. Again, quality should be emphasised, not history.

[1]Actual wording 'After meeting the water quality criteria in the *Australian Drinking Water Guidelines* and the *Australian Guidelines for Water Recycling*, the purified recycled water will be blended into Wivenhoe Dam'.

8.2.4 Microconstituents

In the USA, the WateReuse Association chose 'microconstituents' to replace the numerous cumbersome and intimidating terms such as 'emerging pollutants of concern'. One academic complained that the term 'microconstituents' was not scary enough and would make it more difficult for him to raise funding for his research. An unanswered question about microconstituents is why do researchers concentrate on detecting smaller and smaller amounts of chemicals but do little to determine their effect?

8.3 ANOMALIES

8.3.1 Water reuse is the water industry's best kept secret

Instead of taking advantage of the recycling that is already happening to reassure the public, water reuse is the water industry's best kept secret. There are many examples of recycling along the major rivers of Europe and the United States. Yet water professionals make it seem as if recycling is extraordinary rather than commonplace.

People are not inspired with confidence when they learn that water utilities are responsible for 'inadvertent' and 'unplanned' practices. Why is it all right to recycle water if it is done in an 'unplanned' or 'inadvertent' way but it is not all right to deliberately recycle it, using best practice?

8.3.2 Two sets of guidelines for drinking water

The community is given the impression that planned recycling is more risky than unplanned recycling. In Australia there are two sets of guidelines for drinking water. The *Australian Drinking Water Guidelines* provide for water that comes from conventional sources, including those that contain 'inadvertent' and 'unplanned' effluent discharges. The *Australian Guidelines for Water Recycling: Augmentation of Drinking Water Supplies* are concerned with drinking water that is derived from planned recycling using best practice technology (Figure 8.2). It is not clear why the Guidelines for planned reuse should be different from, and more rigorous than, those for unplanned reuse. It does not give the community confidence in either the quality of their current drinking water or the recycling process.

8.3.3 How long is the miracle mile?

A distinction is made between direct and indirect potable reuse. The process train for indirect reuse includes an environmental barrier, a process whereby clean water is mixed into a dirty dam or allowed to flow down a muddy river for a Miracle Mile. We are told that this will save us from plant breakdowns, inadequately trained or incompetent operators and all other disasters. The theory is based on the outdated principle that 'dilution is the solution to pollution'. Dilution and delay were used in the past to improve the quality of effluent but we now have modern technology to do the job for us, giving us more control.

The amount of dilution and residence time are not specified – perhaps because it would limit the use of the water when it is most needed – when reservoirs are low. There have been few studies to ascertain if an environmental barrier is the best way to spend money and energy to ensure the safety of the water. In Europe, the emphasis has been on providing more treatment at the water treatment plant, such as ozonation and filtration through activated carbon, rather than having sophisticated wastewater treatment and reclamation plants. The irrational insistence on an environmental barrier raises doubts in the minds of the community as it suggests that the water quality is sub-standard so it needs blending and it undermines the role and image of the utility.

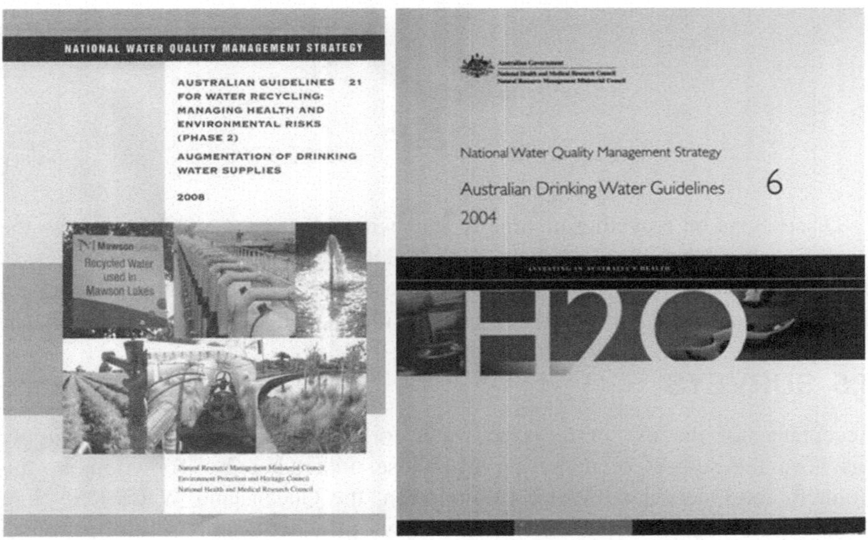

Figure 8.2 Australian Guidelines for Water Recycling and Australian Drinking Water Guidelines

8.4 WE START TOO LATE

Consultation with the community invariably starts only after the decision to recycle has been made – it is DIDI … D (**D**ecide; **I**nform; **D**efend; **I**gnore … **D**uck) – whereas it should start by identifying the problem and all the options available for solving it. Typical of this approach was the proposal to recycle water in Toowoomba where any alternatives strategies to recycling – and their deficiencies – were not convincingly explained to the community. The water industry should be proactive in providing basic information **before** recycling becomes an issue – an education program cannot be retrofitted; the community sees it as 'marketing' and rejects it.

Discussions on recycling invariably start with the processes used to purify the water whereas they need to start with a knowledge of basic water science, particularly, wastewater and how it is treated. If the community does not understand wastewater, they will not be able to understand and accept recycled water.

8.5 SURVEYS

Acceptance of the use of recycled water for augmenting the drinking supply has been the subject of many surveys those undertaken by Toowoomba City Council, the Queensland Water Commission, the Queensland Water Recycling Strategy and many others. Most of them have provided minimal information to the participants. Some noted that those surveyed became less likely to accept recycled water after the survey. The surveys have produced results that are not significantly different. 20 to 25% of people, acceptors, are willing to drink recycled water, 50 to 60%, doubters, have reservations and 20 to 25%, opposers, are unwilling. The results in many cases reflected the terminology used in the questions and any recent activity of those opposed to recycling. Most of the effort focused on what is causing the aversion rather than how to overcome it.

My personnel experience has been that people change their views about recycled water when they have more knowledge and awareness of water. In a local consultation about water management in Noosa on Queensland's Sunshine Coast many people responded to information days organised by the Sunshine Coast Environment Council and community groups. The community consultants later surveyed the community's preferred options. The results showed clearly that the people who had attended information days had a greater understanding of wastewater and its treatment and accepted a wider range of options, including direct potable reuse, compared to those who had not attended. It was also apparent that some people would never accept potable re-use.

	With information	Without information
Indirect potable re-use	0%	7%
Land disposal	11%	22%
Disposal to Creek	33%	32%
Direct potable re-use	56%	39%

This reflects the results obtained by Loretta Lohman, Research Social Scientist, in a survey she undertook to study the attitude of the population of Denver, Colorado, to reclaimed water while their reclamation plant was under construction (Lohman & Milliken, 1985). She discovered that the community could be divided into the three groups: the first group had no problems with the concept of reclaimed water and 'didn't mind at all' even the idea of drinking it. The second group 'minded a little' mainly because they lacked trust in the technology. The third group 'minded a lot' and mostly did not change their attitude that was based on emotion.

Ms. Lohman demonstrated that, with information, familiarisation, reassurance and particularly an on-site inspection of the plant, a high proportion of this group came to accept recycled water. In one group the number of 'don't mind at all's' rose from 9% to 64% after a tour of the plant.

8.6 THE PROBLEM IS NOT BEING ADDRESSED

The conviction of the acceptors is not robust because it is not supported by a sound knowledge of water science; they are unable to reassure the doubters. The doubters are vulnerable to the efforts of the well-funded politically motivated opposers who bombard them with one-sided emotive negativity – so easy!

Water professionals have allowed this situation to develop without doing anything effective to counter it. Researchers appear to have more interest in examining why those against recycling have their opinions than in finding ways in which their irrational and emotional misperceptions may be overcome. The communicators who are responsible for providing information to the community sometimes appear to have an inadequate knowledge of water science. For example, Water Environment Federation in their community targeted publication, *Following the Flow – An Inside Look at Wastewater Treatment*, pharmaceuticals and personnel care products are classified as inorganic chemicals.

Resistance to recycling has continued and intensified because the water industry has not addressed the community's lack of knowledge. If the community is to feel comfortable about the safety of recycled water, they need to know what is put into water, how it is taken out again and how they can be sure that it has been taken out. However with inadequate, inconsistent and frequently stigmatising terminology, confusing anomalies and inadequate attempts to provide information, is it any wonder they are wary?

8.7 THE WAY FORWARD

8.7.1 Terminology

New terminology is needed to define reclaimed water quality by the uses for which it is suitable as opposed to its source and degree of treatment as the present. National (and preferably international) agreement on this is important in order that the community is not confused by a variety of terms. Professional water organisations should take a leading role in this and those responsible for water industry publications should draw attention to inappropriate terminology. Water professionals should be made aware of the impact of the words they use and be persuaded to make an effort to overcome their bad habits.

8.7.2 Training for presenters

Currently the number of university courses that include water science is limited; even courses in environmental issues do not include basic information about urban water and wastewater management. More are needed so that graduates are equipped to explain water science to a lay community. Community consultants must do more to overcome the problem that uninformed people have uninformed opinions. They should either have sufficient knowledge of water to be able to explain the water science themselves or have access to experts who have both the knowledge and presentation skills. Basic water and wastewater management should be part of the high school curriculum.

8.7.3 Cultural changes

Professional water engineering organisations should encourage their members to consider the whole of the water cycle and not parts of it in isolation (Figure 8.3). Conventional water management has been divided into well-delineated sections – water supply and wastewater management – each dealt with separately and kept as far apart from each other as possible, both physically and mentally. This division needs to be overcome if water recycling is to be accepted.

Water professionals must consider that community acceptance of any recycling project to be equally as important as the technical aspects because, if the community's concerns are not addressed, engineers will be continually frustrated in their attempts to manage water sustainably. Community outreach must be generously funded.

8.7.4 Anomalies

There are several regulatory anomalies that need to be resolved if the community is going to be able to accept replenishing the drinking supply with recycled water. The emphasis on the need for an environmental barrier that suggests there is something wrong with the water, the irrational distinction between planned and unplanned potable recycling and the two sets of guidelines for the same product, drinking water, are scarcely reassuring to the public. As the community becomes increasingly informed they will question these aberrations.

8.7.5 Establishing trust

In order that the community may trust recycled water it is important that they trust those who are providing it for them, the utilities. Water is a manufactured product and the utility should be seen as a provider of its quality as opposed to the source providing it. Utilities should endeavour to distance themselves as much as possible from 'government' as, in general, the community does not trust 'government'. They should use marketing tools cautiously as attempts to 'sell' water reclamation to the community are frequently and cynically seen as 'spin'.

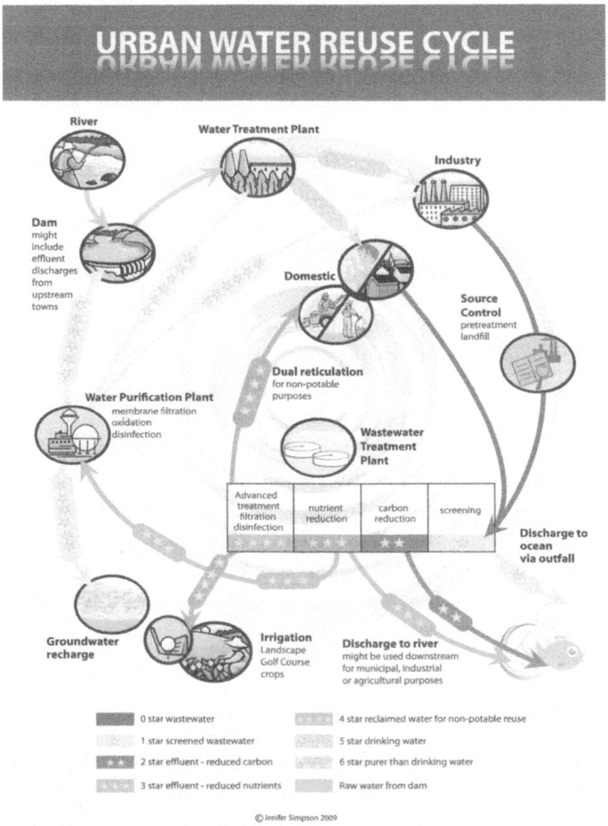

Figure 8.3 Urban Water Reuse Cycle

8.8 AN EDUCATION PROGRAM IS URGENTLY NEEDED

Many stakeholders do not want detailed information but information should be made available for those who do want it. Those who want the information frequently have a strong role in positively influencing others and the knowledge that the information is available is reassuring to all.

There needs to be an **active** water science education program that simplifies the concepts and uses easy to understand words and images. It should be **imaginative, stimulating** and cater for all learning preferences and styles. It should be independent of the media, **energetically and charismatically delivered to the community** and take advantage of modern communication techniques. It should not rely on people being sufficiently motivated to visit websites. It would include, not be confined to, mobile and static exhibitions, interpretive centres as well as creative materials distributed by all methods of communication such as websites, brochures.

This diagram encourages discussion of the total urban water reuse cycle instead of bits of it in isolation as invariably is the case with community targeted information. It illustrates the star rating system to explain water quality as described in the author's booklet, *From Waste-d-water to pure water*, winner of the 2008 IWA award for 'Best popular presentation of water science'. Over 20,000 of these books have been circulated in Australia and a further 10,000 recently been printed by WaterSecure as part of their water education program.

REFERENCES

Lohman, L.C., and J.G. Milliken, "Informational/Educational Approaches to Public Attitutes on Potable Reuse of Wastewater", University of Denver Research Insitute, August 1985 (available from NTIS).

Chapter 9

Water: nature's amazing reusable resource

Linda Macpherson
CH2M HILL, Portland, Oregon, USA

ABSTRACT

The notion of reused water can evoke negative images. The ensuing discourse can rapidly degenerate into a process of rationalizing fears that are based upon initial perceptions of the stigma that used water may have gained from the language and images used to describe it, and often results in failed implementation of much needed water reuse projects. While much of this existing attitudinal landscape is beyond the control of water managers, some of it is exacerbated by the vocabulary and images typically used when defining and educating our population about water, the water cycle, and water and wastewater treatment. By developing a clear understanding of these interlinked elements and processes through the use of positive, emotionally engaging, understandable words, images, and experiences, water managers and communicators can overcome the stigma of water reuse that is based on fear, not fact, and deliver successfully supported water reuse projects that become proud fixtures of our communities for generations to come.

This chapter discusses why water reuse projects have failed to gain the needed community support to make them reality, including political and media involvement, a history of a linear approach to water management, and a poor understanding of water in general by the public. as well Key components of successful reuse project outreach, including proper terminology, imagery, experiences, visitor centers, and other real-life tactics for success are also described. This chapter provides guidance on how to approach educational outreach on water reuse in a sustainable manner so that generations to come continue to support, and even value, water reuse as a means to a healthy community, environment, and economy.

Keywords: Education; emotion; perception; sustainable; stigma; water reclamation; water reuse; outreach

9.1 INTRODUCTION

The notion of reused water can evoke negative images. The ensuing discourse can rapidly degenerate into a process of rationalizing fears that are based upon initial perceptions of the stigma that used water may have gained from the language and images used to describe it. A further complication is the public's general lack of understanding about the water cycle. Couple this with a healthy dose of public skepticism and distrust of technology, and a needed water supply strategy can easily become doomed. While much of this existing attitudinal landscape is beyond the control of water managers, some of it is exacerbated by the vocabulary and images typically used when defining and educating our population about water, the water cycle, and water and wastewater treatment. By developing a clear understanding of these interlinked elements and processes through the use of positive, emotionally engaging, understandable words, images, and experiences, water managers and communicators can overcome the stigma of water reuse that is based on fear, not fact, and deliver successfully supported water reuse projects that become proud fixtures of our communities for generations to come.

The water and wastewater industry's relationship with the public has been a turbulent affair. In the 1930s its entire existence was kept out of sight and out of mind, and even when the world population reached 2.5 billion and a strong push for wastewater treatment occurred in the 1950s, facilities were built outside of town where the less said the better. In 1955 pop culture even made fun of wastewater operators as Art Carney sang the 'Song of the Sewer.' In 1975 Paris openly embraced and celebrated the wastewater industry and the engineers who developed and continue to modernize the extensive sewer and water supply networks by opening the official Paris Sewers Museum and leading The Sewer Tour. By the 1990s, public visibility of the water industry had grown, and wastewater treatment plants used innovations in design and engineering to ensure no odor and to add positive qualities to neighborhoods, including beautiful architecture and landscaping. By the 2000s, many of the newly designed and expanded treatment plants were designed to attract visitors, be community assets, and showcase the treatment technology to promote understanding and acceptance of water reuse. While there are a growing number of wastewater and water reuse success stories, a large number of projects continue to fail to gain the needed support to turn renderings into reality. New approaches to understanding the urban water cycle are needed.

Importantly, new water resource management solutions like water reuse are needed to deal with current and anticipated water stress caused by population growth and rapid industrialization, drought and climate change, crisis in supply (the easy sources of water have already been tapped), and mental models that demand a source that is thought to be 'untouched,' 'unused,' and from a 'pure' source. The need is critical because we have an insatiable thirst for water, driven by rapid population growth and by the number of people living closer and closer together in our cities. A system that evolved for a population that was mostly rural and without modern technology will not be the solution that is needed when we have a global population of 9 billion in 2040 (mostly urban), and are experiencing greater resource constraints and climate change. It is no longer possible to be 'out of sight/out of mind'. Much of the urbanized world has the knowledge, technology, and resources to implement projects that will help us to maintain a sustainable water supply, but without community support these projects, and consequently our economies, health and futures, are in real danger. It is time to make the invisible–visible.

9.2 WHY REUSE PROJECTS FAIL

In order to design successful public outreach and education programs that overcome public stigma and fear of water reuse, it is beneficial to first look at why water reuse projects in the past have failed. Many in the water business believe that the public's lack of knowledge about water and the water cycle in general is the single largest barrier to sustainable water management. This lack of knowledge starts at an early age, as most of us learn about the water cycle as though there were no human interface. We forget that the water cycle and the life cycle are one. In truth, water has been, and always will be, nature's amazing reusable resource.

- The Earth has always had the same amount of water
- What we have now is all we are ever going to have
- The water we drink has been around for 3.8 billion years
- We are living closer and closer together as population increases
- My water becomes your water becomes their water all around the world
- We already reuse water!

Water reuse is nothing new—we have been doing it for centuries—the reality is global water reuse is the water industry's best kept secret. When we focus on recycled water for drinking, we consider only planned indirect potable use, but in truth, unplanned, inadvertent, defacto indirect potable use has occurred for decades in many parts of the industrialized world with *no known health issues.* For example, 360 wastewater treatment plants discharge into the Thames River and nearly 7.3 million customers consume the water daily as their source of drinking water. Wastewater treatment plants also discharge into the Rhine, which is a source of drinking water for 20 million people including those in Bonn, Cologne, Rotterdam, and the Netherlands. More than 80 wastewater treatment plants discharge treated effluent into New York City's three drinking water sources, yet no increase in health risk has ever been identified. In fact, a very large part of the population of the world's industrialized countries practice indirect potable reuse, but we seem to only focus on it when it is *our* water intake that is below *our* discharge. This is a result of not putting the subject of reuse in context—making it sound extraordinary rather than commonplace; *we are all downstream*.

Additionally, the water and wastewater industry has disproportionately focused on the history of water, rather than on its treatment and resulting qualities (and, hence, on the ultimate safety for its intended use). In some instances, we feel it is critical that people know that the water was once used, but we tend to focus much more on what water once was than on the means that engineers and scientists have to treat and make water ready for its return to the water cycle.

The compartmentalization of water supply and wastewater treatment into different organizations has added to this poor understanding of the water-cycle and how it is managed in context with human interaction. The idea of 'take, make, waste,' has a particularly negative impact on water reuse, especially where water and sanitation services are run by different entities. When we think of our system as linear, with treated water discharged into the waters of amnesia, and then appropriately define water reuse as unique, we make the usual sound unusual — and we know that we fear things that are out of the ordinary and seem strange. Sustainable approaches are needed that define sustainability as the principle of optimizing the benefits of a system without diminishing the capacity for similar benefits in the future. The way our industry currently defines reuse leads to unsustainable decisions that are based on fear, not fact.

We rarely explain 'water quality' and how we 'treat' water using science, clear language, and real-life examples prior to outreach about a specific reuse project. Discussions of reuse and water issues in general typically don't start until *after* a community has already begun to look at their options for a sustainable water future. If the public has a limited understanding of water issues, the water cycle, or the basic information about the chemistry and microbiology of water, then they are less likely to understand water quality and water treatment and be able to effectively weigh the benefits and tradeoffs of various treatment alternatives.

When having to choose between something that is familiar and something that is not understood—and even openly feared and condemned—people tend to go with the familiar. In general, educational outreach, as well as the engagement of vital community opinion leaders, starts just too late in the process, allowing for negative campaigns to easily spread and build fear and stigma within the community.

Reuse decisions can, and often do, become a political football, and politicians are known to take advantage of emotive images or phrases to defeat projects or create stigma. For instance, in the late 1990s in California, The San Diego City Council and the San Diego County Water Authority commenced a 'water repurification' project that seemed to have all of the public outreach and information programs in place for public acceptance success. Conceived during a drought, the recycled water would be mixed with imported 'freshwater' in reservoirs to supplement the city's drinking water supplies. Focus groups, interviews, and telephone surveys found that residents would be supportive of using the repurified water for household and potable uses. However, despite the strong support from numerous community groups and public figures, the project became entangled in political campaigns that highlighted health dangers and claimed that the city intended to take wastewater from affluent communities to distribute as drinking water to those less affluent. Campaign posters with the slogan 'Toilet to Tap' also focused the public's attention on the source of the water rather than the process and final quality, causing the project to be put indefinitely on hold. The City of San Diego has since built two water reclamation plants, but the treated water is approved for irrigation, manufacturing and other non-drinking or non potable purposes only. The City of San Diego is currently undertaking a Demonstration Advanced Water Purification Facility with the goal of showcasing the technology in a more open and transparent fashion. It is theorized that this 'seeing is believing' approach, placed in a water cycle context of use and reuse in San Diego and the world, may hold a different result.

To combat easily spread fear and stigma we have to do a better job of using language and imagery to create a deeper understanding of how we use water and how we learn about treating water. Currently, reuse project communications typically rely on technical analysis, fact sheets, and explanatory meetings that are characterized by reason, logic, and the use of technical terms to describe advanced technology and water treatment, all of which can be off-putting to, or not understood by, the community at large. We say that water is 'dirty' when chemicals, silt and other materials moves through the water cycle; industry terminology, such as treated sewage, sewage treatment plant, treated wastewater, and disposal, while somewhat accurate to an engineering mind, elicit negative emotions and mental imagery for the nontechnical stakeholder so that no logical or rational argument can easily overcome. As discussed in Sadie Dingfelder's 2004 *Monitor* article 'From toilet to tap,' Brent Haddad, PhD, an associate professor of environmental studies at the University of California, Santa Cruz who studies the success of water reuse initiatives, and Paul Rozin, PhD, a University of Pennsylvania psychologist who studies disgust and contamination, find that this emotional opposition comes more from a knee-jerk response to wastewater—the yuck factor—than from concerns about the water's chemical composition. This emotional factor can be changed by creating settings for water learning in which the learner is exposed to images and ideas where they can chose what, how, where and with whom to learn the information, rather than instances, such as public meetings, where stanch opponents use terminology that associates the quality of the water with wastewater.

9.3 KEY COMPONENTS OF SUCCESSFUL REUSE OUTREACH PROGRAMS

It is critical to work through these issues in a manner that encompasses rational thought as well as emotion and feelings. When members of the general public are given the opportunity to reframe their understanding about water and to learn and participate with a more realistic view of water quality, they draw different conclusions. A number of countries and cities around the world are pointing the way to developing sustainable solutions and are suggesting steps for creating a better understanding of water use and reuse. These approaches can help to facilitate deeper acceptance and ownership of the water cycle and water treatment.

Informed decisions require informed citizens. Citizens need to be informed about all water supply options (reuse is but one of the options) so that they can compare the cost, reliability, energy consumption, social disruptions, and environmental impact of all options. The public needs to understand water quality and water treatment; education, polling and votes about recycling alone will not suffice and serves to stigmatize. Overcoming negative perceptions is possible by providing knowledge and understanding of water quality and treatment and by focusing on the quality of the water, not its source. By acknowledging where and how water reuse already happens around the world we can make reuse 'normal' and remove the fear of the unusual. For this reason, the first step to consider when planning a reuse/recycling project is to gauge whether public understanding of water issues is cloudy or clear before proceeding with a suggested project. Developing a water supply planning program *with* the public that educates and helps them to understand the issues and tradeoffs, benefits and consequences, the chemistry and microbiology water, the water cycle and reuse around the water, will establish trust and will lead to *reasoned* decisions when it comes time to discuss the tradeoffs of possible solutions. It is vital to start these communications and build understanding before the specific reuse project is introduced, and to conscientiously engage the media ahead of time to be on our side and to ensure unbiased and fair reporting. For instance, to reduce scepticism by the end users regarding water quality, Singapore PUB worked well ahead of delivery to educate the potential end users of the NEWater plant, a water-reclamation facility that is part of the 4 National Taps program to supply drinking water to the nearly 4 million citizens. Keeping the community fully informed about the level of control and redundancy being provided to ensure the plant and system reliability in terms of production and quality is also critical. PUB also engaged the media, key public, government, and community figures from the very beginning to rally support and understanding for the project. PUB showed the media use and reuse projects around the world.

As education and knowledge of water are important to establish trust, it is important that community opinion leaders be approached early, as it will increase the prospects for *rational* positions, whether for or against. As many citizens make decisions based on the opinion of trusted members of the community, those who look for leadership and guidance from these leaders will also benefit. For example, PUB bottled NEWater in attractive packaging so that the public could sample for themselves how pure the water is, and these were distributed at grassroots and national events. Top government officials including the then Prime Minister of Singapore became NEWater ambassadors and champions when they were seen drinking NEWater publicly. This was a huge boost to the campaign, as it positioned the project as a national effort led by the country's leaders. During the 2002 National Day, some 60,000 Singaporeans toasted to NEWater in a show of solidarity demonstrating the support and confidence they had in NEWater.

Community trust in the government or utility is also critical. This trust can be developed through transparency of communications, education, community investment, and creating a forum, physical or virtual, where the public can express and have their concerns and fears addressed in an understandable, positive, emotionally engaging manner. Having community leaders, and citizens, engage with and hear testimonials from leaders and citizens of communities who currently use water reuse also proves to be beneficial as it creates familiarity and security that the technology isn't unusual and is safe.

As we recognize that inadequate and negative terminology impedes clear communication, we can work to create programs that use images and words to enhance knowledge and understanding of water (See Figure 9.1 and 9.2). How we put things in water and take them out again must be successfully conveyed to the public or fear and phobia will take over and lead to another failed reuse project. Because perceptions have such impact, we all must stop using negative terminology that stigmatizes. Use 'reclaimed water' or 'purified water' rather than 'treated sewage'; 'water reclamation facility' rather than 'sewage treatment plant'; 'renewed' or 'purified' water rather than 'treated wastewater'; 'recycling' rather than 'disposal'. Our capacity to absorb factual information is blocked by negative associations of 'dirty' water, so we must make an effort to use terms that are accurate and positively resonate with the public, like sustainability, drought-proof, safe, or secure. This terminology helps people to understand that the benefit of water reuse recycling technologies is not just a future, but *our sustainable future*.

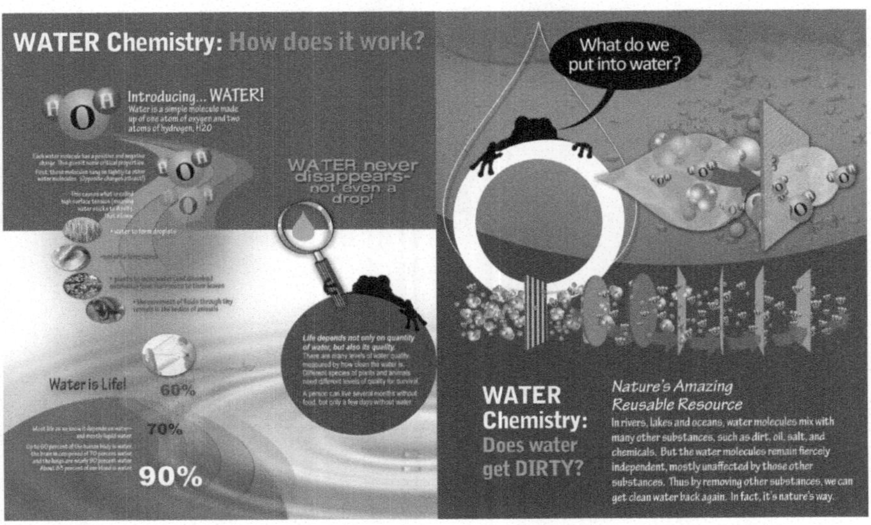

Figure 9.1 Materials that explain water basics should be used early in the outreach campaign, before specific project solutions are ever discussed, so that citizens have the understanding to effectively weigh the benefits and tradeoffs of treatment alternatives

Figure 9.2 Educating and engaging both citizens and the media on water and proper terminology as early as possible in the project planning process can help to prevent media campaigns like these from creating fear of, and further stigmatizing, water reuse

Rather than the highly technical analysis, fact sheets and explanatory meetings, understandable, citizen relevant images and terminology that create positive feelings that resonate should make up the project's communication outreach (See Figure 9.3). Citizens are feeling human beings, and we can no longer design and present our outreach programs as if they were purely rational, logical, un-feeling creatures.

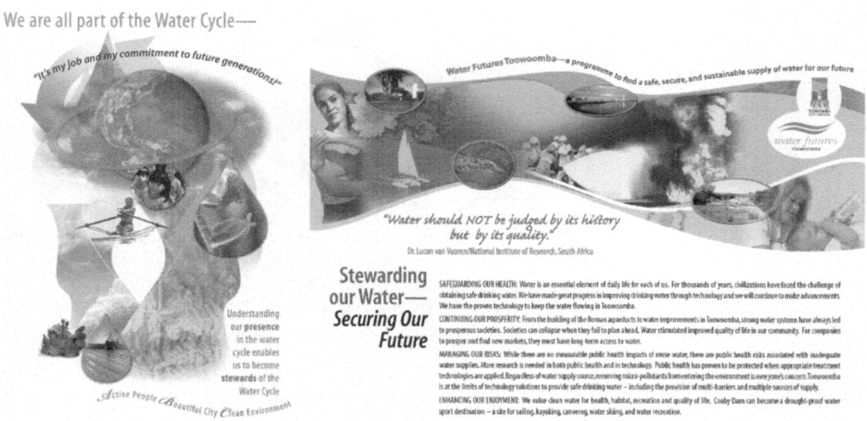

Figure 9.3 People centric images that reinforce health, safety, happiness, family, prosperity, positive interaction with the environment and water, and are culturally familiar engage viewers in a positive manner

Turning highly technical information and graphics into understandable processes by using familiar everyday objects and processes is also useful, as seen in Figure 9.4. At a time when water is now recognised as a precious and limited resource, the Gippsland Water Factory is a project the local community can be proud as it will not only generate recycled water for local industry and significantly reduce the offensive odour emitted from the open channel section of the outfall sewer, but also will save about three billion litres of fresh water each year. Gippsland Water drew attention to these benefits in their outreach and planning process, as well as how these advances will highlight the region as a leader in sustainability and innovation, creating community pride, warm feelings, and support for the project, as seen in Figure 9.5.

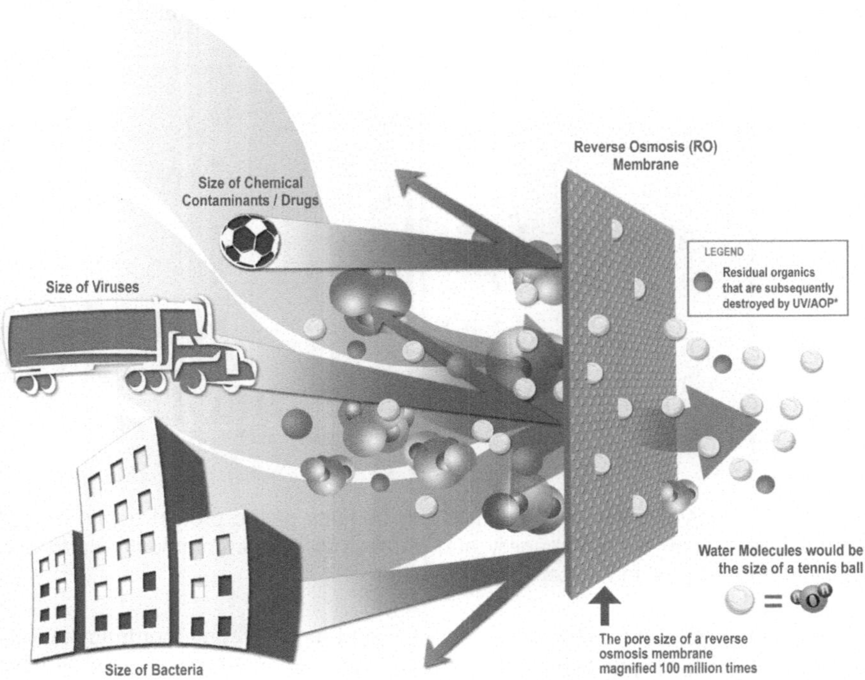

Figure 9.4 Familiar sized objects and images are used to describe the reverse osmosis process in an understandable manner, preventing fear and skepticism that highly technical explanations of advanced technology can create

Rozin's studies have also found that reuse advocates and public officials can counteract perceived contamination by associating reclaimed wastewater with environmental groups, such as the Sierra Club, through advertisers and endorsements. Rozin suggests that by capitalizing on this principle, water managers can effectively neutralize the liquid's perceived link to human waste. As the greater the benefit is perceived the lower the perceived risk, highlighting benefits and developing a strong understanding that reclaimed water is inextricably linked with economic success and a sustainable future will also help to overcome public outrage. This method was used successfully in Singapore's award-winning public information strategy that occurred prior to opening the NEWater factory at Bedock that included a visitor centre. PUB publicized the importance of reclaimed water to the continued success of the country's economy and focused on how the new water source would significantly decrease the country's dependence just one source of water. The importance with punctuated in the tag line – NEWater: For Industry, For Living and For Life. The tagline was repeated in the videos and other experiences inside the visitor centre.

Figure 9.5 Gippsland Water highlighted the benefits of the project using familiar and positive human centric imagery and phrases to create community pride and support for the Gippsland Water Factory

New education and outreach techniques and non-meeting communications capture attention, expand information dissemination, and limit conflict. By providing innovative ways for people to communicate about reuse, through eCommunication, eVolvement, and other social media tools, the water industry can expand stakeholder understanding and even affect behavior change. Social media and marketing not only help sell the benefits of behaviour/attitudinal change, but also actually encourage the community to help create and evolve the change themselves. Social media allows for success stories, from both their own community and others who have successfully adopted water reuse, to be shared amongst the community, spurring on others to make similar behaviour changes as they see the positive results and reinforcement of alternative behaviours. Using the internet and a dedicated website and e-tools throughout the project can help users explore the issues and expand the conversation beyond the merely stigmatizing statements and images, can let the public see the issues in new ways and decide for themselves, as well as can minimize the public's need for internet searches, which frequently result in taking statements out of context and creating distorted meaning. Through proactive blogs and going beyond meetings through the use of video sharing and live online Q&As, water managers can facilitate transparent communication and build trust. Conversations can occur between citizens and the governments, between citizens of the same community, and even citizens in different communities. These resources will stay and grow with the community as the project is delivered, keeping the public up to date on key events, project progress, and activities that might affect businesses and homeowners. It also allows the government and water managers to monitor conversations and correct misinformation.

9.4 DEVELOPING SUSTAINABLE COMMUNITY SUPPORT

In addition to a successful community outreach and education program that occurs prior to and during project development, it is vital that continued communications and positive experiences with water reuse and associated technology continue, especially as communities grow and expand. In addition to traditional public outreach campaigns, the development of visitor learning and demonstration centers at the reclamation facilities are proving to be the most successful method for developing the long-term acceptance and ownership of water reuse. The visitor centers are places of discovery and learning where people come to understand that water, climate, economy and future are interlinked; where the journey of water becomes integrated with a map to the future of the community. These visitor centers can include water cycle references, compelling architecture, interactive displays and activities, a narrative journey, exposure to the treatment plant's working technology and could even be linked to each other.

The Gippsland Water Factory is the first-of-its-kind in Australia, highlighting Gippsland as a leader in sustainability and innovation by delivering a range of benefits for the region. In addition, electricity is produced onsite using renewable energy technology, which helps power operational activities and minimise environmental impacts on the site and throughout the region. A public education facility, the 'Vortex Centre' raises community awareness about water conservation and sustainable water management (See Figure 9.6).

The state-of-the-art 'green' facility features interactive displays, touch screens and videos, with a focus on water conservation and sustainable water management, highlighting water as a precious resource at a local, state, national and global level. As visitors move through the centre, which is shaped as a vortex to symbolize the integration of water, energy, climate, economy, and the community's future, they learn about the water cycle, Gippsland's water catchment and how it is managed, the Gippsland Water Factory treatment process, and how we can all play a part in conserving water and using it more wisely. The Vortex Centre is free for those who live within the service region, and school groups are encouraged to bring students to ensure the next generation is well-educated on water. Even though water is emphasized, every opportunity is taken to deepen the visitors' connection to the way water influences the Gippsland region's natural and economic systems. The goal is to help visitors understand how we affect water, how water is a part of our livelihood, how sustainability enables us to secure a positive future, and ultimately, how being a better steward to our planet is being a better steward to ourselves.

Figure 9.6 Left: The Gippsland Water Factory Vortex Centre was built adjacent to the Gippsland Water Factory so that visitors get first-hand experience of the technology treating their water. Right: Inside the Vortex Centre

Central to the entire public education for Singapore's NEWater campaign was the setting up of the NEWater Visitor Centre, a state-of-the-art water museum that acts as a one-stop centre for anyone looking to understand how NEWater is produced and how it plays a part in Singapore's water strategy (See Figure 9.7). This helped, and continues to help, visitors to understand the introduction of NEWater in the context of Singapore's water story. Great pain was taken to ensure that the information presented at the center is palatable and easily understood by the public. Technical information on the technology was broken down into layman terms using multimedia animations and interactive touch-screens. For instance, to enable the younger generation to fully understand the process of reverse osmosis, familiar objects like ping pong balls and tennis balls were used to help illustrate how the process works.

To enhance the learning experience for visitors, the NEWater plant was incorporated as part of the centre where visitors could actually see the membranes used in the purification process. An independent survey by Forbes Research at the end of 2002 confirmed the success of the programme. NEWater had garnered a 98 per cent acceptance rate, with 82 per cent of respondents indicating that they would drink NEWater directly and another 16 per cent would drink it mixed with reservoir water. Confidence building is not a one-off process and instead of calling the campaign successful and complete, PUB continued to sustain its public education campaign. This was critical to avoid being subjected to the 'out of sight, out of mind' mentality by the public who may forget why and how they came to accept NEWater. The NEWater Visitor Centre continues to be the main vehicle for sustained public education on NEWater. Key milestones such as the opening of the fourth NEWater plant, commemoration of the 1st and 5th millionth NEWater bottles were publicized to keep NEWater in the public eye and more importantly, maintain the public's confidence in NEWater. Singapore continues to invest in providing spaces that perform needed functions but are also fun and enhance learning.

Figure 9.7 Left: The NEWater Visitor Centre engages visitors with the water, environment, and treatment process before even entering the facility. Right: Inside the NEWater Visitor Centre visitors learn about Singapore's water and the treatment through interactive displays and the 'Liquid Life' narrative

Improving water education in our school systems so that the water cycle includes human involvement and water and wastewater treatment is also a key to building sustainable support for water reuse, as the next generation will then already find the technology usual, rather than unusual. The current approach showing the natural water cycle of precipitation, transpiration and condensation does not show the urban cycle. By bringing the water cycle and the life cycle together, we can support the young, and consequently their parents, in thinking about water use and reuse from a whole new perspective.

9.5 THE WORLD WON'T THINK DIFFERENTLY ABOUT WATER UNTIL THE WATER INDUSTRY STARTS THINKING DIFFERENTLY

The vision of sustainable water management will not be achieved through conventional thinking in the silos defined by different technical water disciplines.

It will fail because of the linear 'take, make, waste' thinking that characterizes our wastewater as disposable, rather than as a usable and reusable resource. This approach leads us to think that water is something we use, but it goes away (out of sight and out of mind), rather than something we can use and reuse — water made usable through technology.

As Bruce Durham, technical secretary for the European Federation of National Associations of Water and Wastewater Services' water reuse working group, so elegantly put it, 'The 'clean water suppliers' and the 'others who treat wastewater' can no longer remain at arms length; they have to accept that they are dealing with a single resource that cycles through a complex of quality phases, all of them interlinked.'

Real paradigm change will occur only when we create a holistic, integrated, and multi-disciplinary perspective and approach to water and the water environment. Solutions must be found through systems thinking that captures both hearts and minds. It requires engineers, scientists, sociologists, and educators working collaboratively. It requires an agreement to invest in communication innovations along with those in science and technology.

Water is, and always has been, nature's amazing reusable resource. Public understanding and support are keys to sustainable water management. We need to remove barriers to understanding by changing and eliminating stigmatizing language and by creating interesting learning opportunities that take water management out of the closet and that make the invisible, visible. Many water industry leaders are building education centers and creating community and civic assets that are yielding positive dividends. The result is enhanced trust in the people, technology, and expertise of the water industry. To move toward sustainable water management, we need to show people that we provide safe and socially responsible services. We need to tell people what we put into water and how we take it out again. We need an educational approach that focuses on quality (not just the source). And we need clear and compelling education programs.

Water ties us to our history and enables our future. It is a primary ingredient in us and all we do. Its quality and use connect the future for us all. Sustainable water management depends on opening people's hearts and informing their minds. Can the change come from the water industry? It must.

REFERENCES

Daigger, G.T. (2008). *Evolving Urban Water and Residuals Management Paradigms: Water Reclamation and Reuse, Decentralization, Resource Recovery*. AEESP Lecture, Water Environment Federation 81st Annual Conference and Exposition, WEFTEC '08.

Dingfelder, S. (2004). From toilet to tap: Psychologists lend their expertise to overcoming the public's aversion to reclaimed water. *Monitor*. Volume 35, No.8.

Faldenmark, M. and Lindh, G. (1993). *Water and economic development, 'Water in crisis.'* Pacific Institute for Studies in Development, Environment and Security, Stockholm Environmental Institute.

Lohman, L.C. (1987). "Potable Wastewater Reuse Can Win Public Support." In *Proceedings of Water Reuse Symposium IV*, Denver, Colorado, August 2–7, 1987. Published by the AWWA Research Foundation, Denver Colorado.

Roseth, N. (2008). *Community Views on Recycled Water – The Impact of Information.* Cooperative Research Centre for Water Quality and Treatment 2008, Research Report 48.

Slovic, P. (2000). *The perception of risk.* Earthscan Publications, London, United Kingdom.

Sunstein, C.R. (2002). Risk and Reason: Safety Law and the Environment.

Kluger, J. (December 4, 2006). Why We Worry about the Wrong Things; The Psychology of Risk. Time Magazine.

Chapter 10

From zero to hero: NEWater wins public confidence in Singapore

Yap Kheng Guan and Sally Toh
Singapore PUB, Singapore

ABSTRACT

The crux of a successful water recycling programme lies not in assembling the best technology available or ensuring the quality of the water, but in winning public confidence and acceptance. This was amply demonstrated by PUB, Singapore's national water agency, with the introduction of its own brand of reclaimed water 'NEWater' in 2002. Its success in achieving widespread public acceptance sets PUB's NEWater programme as a pioneer in large-scale water recycling and more importantly, as the international beacon for public education on reclaimed water.

From the outset, PUB designed a comprehensive public communications programme to engage the various groups of stakeholders including a familiarization trip to the USA for media, briefings for Members of Parliament, grassroots leaders and business groups, a panel of experts who addressed health, safety and quality issues and the setting up of the NEWater Visitor Centre. This paper gives the history and on-going efforts of the SIngapore PUB to pro-actively engaged with its citizens and the media to ensure the success of their NEWater product.

Keywords: water recycling; reclaimed water; education; NEWater; communications

10.1 BACKGROUND

Although not naturally endowed with an abundance of land or water, Singapore today is recognized internationally for its effective water management system. PUB, the national water agency of Singapore was conferred the Water Agency of the Year award at the Global Water Awards organized by Global Water Intelligence in 2006. A year later, PUB had the honour of winning the Stockholm Industry Water award, one of the most prestigious international awards in the global water arena.

To ensure a robust and sustainable supply of water for its people, PUB had put in place a long-term water supply strategy known as the Four National Taps, comprising water from four different sources – water from local catchments, imported water, NEWater (Singapore's own brand of reclaimed water) and desalinated water. As of 2010, the combined capacity of Singapore's five NEWater plants could meet 30% of Singapore's water supply while desalinated water could meet 10%. Complementing the supply side strategy is a comprehensive water demand management programme that encourages the public to play their part in using water wisely, keeping the catchments and waterways clean and enjoying the recreational opportunities these water resources bring.

10.2 WATER RECYCLING: BANE OR BOON?

Water recycling is a subject that has been much explored and discussed in recent years, especially at a time when the world is grappling with issues like climate change and water pollution. In theory, it looks like the perfect solution to the global water crisis. By allowing every drop of water to be used and re-used, the cycle can go on indefinitely, thereby multiplying the water supply many times over. By leveraging technology as its key enabler, water recycling is a more reliable source of water supply than conventional water sources which rely mostly on the rain. With the cost of membrane technology coming down greatly in the last decade, recycled water has become a cost-effective and sustainable source of water supply for cities around the world.

Despite these benefits, why are cities not embracing water recycling as an option to their dwindling water supplies? The answer, it appears, has nothing to do with engineering or technology, but everything to do with public acceptance, an aspect that has unfortunately been vastly neglected and undermined. In fact, the failure to gain public acceptance and support for recycled water has derailed some of the best programmes in the world, and caused water recycling to be aborted in favour of other, often more expensive, alternatives. In San Diego, California, and Tampa Bay, Florida, officials called off plans to implement water recycling after they faced public opposition and outcry over its use. Although both projects were well-designed, with sound engineering principles and an extensive battery of tests to ensure a superior water quality, they could not escape the onslaught of negative public sentiments.

Both the San Diego and Tampa Bay case studies underscore the importance of public opinion for the successful implementation of a water recycling programme. For ages, water managers have only been concerned with getting the technology and engineering right to the exclusion of all other factors, but it is now acknowledged that public confidence and acceptance is what will make or break their programme. Central to this is the role of the media in shaping public perception. Headlines like 'From the toilet to the tap' only serve to generate public opposition to the use of reclaimed water by playing on the public's psyche and emotions. The use of such visual imagery creates stigma and disgust, which interferes with logical reasoning and clouds public perception but more importantly, it erodes public confidence, swinging public opinion in the wrong direction.

10.3 BREAKING THE PSYCHOLOGICAL BARRIER

This was foremost on the mind of PUB, the national water agency of Singapore when it launched its public education campaign in 2002 to acquaint the Singapore public with NEWater, Singapore's own brand of reclaimed water. The introduction of a new source of water was unprecedented for a country that has relied on two traditional sources of water – water from local catchments and imported water – since its independence 40 years ago. How would the public take to this new source of water? Is it safe for drinking? What are the health effects? Would they accept or veto it? These were some of the questions that PUB had to bear in mind when designing the public communications programme. A successful implementation would not only mark a milestone for Singapore's water supply, it would also put Singapore on the world map for water reclamation.

From the outset, PUB was clear about its objective – to get the public to accept NEWater, something which they have not heard of, and embrace it as a source of water supply for Singapore. To do that, PUB had to educate and convince the public on the safety and purity of NEWater. This was not an easy task, as they were basically starting from ground zero. It was like launching a new product that was virtually unheard of, and had no track record. It was thus crucial to address all the health, safety and quality concerns that the public might have, get them to understand the concept and technology, and develop the same kind of confidence they had in PUB water. Equally important was the need to frame the development of NEWater in the right context so as to achieve a common understanding of its role and importance in Singapore's long-term water supply strategy.

The biggest and most difficult challenge however, was how to overturn the public's psychological fear and stigma, as NEWater was going to be a part of daily life, albeit indirectly. Because of its ultra-clean quality it was decided that NEWater would be supplied mainly to industrial and commercial customers especially those that employ high-end processes such as wafer fabrication. A small percentage would be blended with raw reservoir water before being treated for the drinking water supply - termed 'indirect potable use'.

The key to overcoming the psychological barrier is in shifting attention away from the source by focusing on the treatment process, which involves using advanced, state-of-the-art membrane technology. On paper, this looked easy. Singapore had employed some of the very best technology available, putting in place a stringent, multi-barrier process for the production of NEWater. The system included microfiltration, reverse osmosis and ultraviolet disinfection. The challenge was how to translate this technical information in a simple, layman manner so that even a five-year-old would understand, and yet not trivalise the capabilities of the technology and process. It was also pivotal to emphasize the advanced technology employed in the treatment process consistently across all communications platforms.

10.4 CHANGING MINDSETS AND PERCEPTION

The terminology was also a problem. Despite being internationally recognised terms, words like wastewater or sewage have a negative connotation that would only remind people of the source and add to the psychological aversion. The choice of such words can rapidly stigmatise and inappropriately magnify or invite disproportionate concern, as Macpherson and Slovic (2007) warned. A conscious and deliberate attempt was made to re-name the terms that had a negative connotation with terms that would better reflect the process or its value as a resource. Wastewater and sewage were referred to as used water. This also better reflects their true value and significance in the water cycle as water that can be used and re-used, similar to how water recycles itself in nature, and therefore technically not wastewater to be thrown away anymore. The plants were renamed from sewerage treatment plants to water reclamation plants as they were not merely treating the sewage, but now part of the process that reclaims the used water for further reuse.

A mindset change was also needed to get the public to understand how the water cycle happens in nature and how water recycling is a part of this natural process, and therefore, nothing new. In fact, such unplanned indirect potable use – where treated water is discharged into the rivers and recycled for use by the community downstream – has been practised by cities in Europe and around the world for centuries. This is however, not common knowledge to the man in the street who is unaware that the water has been recycled. On the other hand, in planned indirect potable use, water managers plan how the wastewater will be purified and injected into the drinking water supply. The technology used is typically far more advanced than in conventional wastewater treatment plants and is in fact a process that offers more reliability and quality assurance. It is thus critical to explain these two concepts in a clear and succinct way to show that water recycling is but a part of the natural water cycle.

10.5 ENGAGING THE STAKEHOLDERS

Key to the campaign to win public confidence and acceptance boils down to the way organisations relate and communicate with their various groups of stakeholders. This must be done in a meaningful, thoughtful and trustworthy manner, as Macpherson and Slovic (2007) pointed out, so that the public can see the situation and its issues in context and are therefore able to consider the points brought across by advocates and opponents. Engaging the media as partners in the communications campaign is critical to this process as they are the watchdog for the community and often set out to question the authorities. A sound media plan was needed in Singapore to get the media to understand and support the project as something that the country needed to make its water supply system more resilient and sustainable for the future generations.

Prior to the launch of NEWater, the media were briefed and brought on a familiarization trip to places in the United States such as Orange County, California and Scottsdale, Arizona to show them that water recycling is not a new phenomenon, and that it has actually been a way of life for these people for many years. News features that expounded on the use of reclaimed water in these cities ('Tried and tested in US', The Straits Times, 18 July 2002, 'US residents drinking reclaimed water for 27 years', Channel NewsAsia, 17 July 2002) helped the public to understand that this was proven technology that had been around for years. A television documentary 'Thirsting for new sources' which looked at the NEWater technology and the experience of USA in water reuse was aired on local television. In tandem with the media engagement plan, briefings were organised for Members of Parliament, grassroots leaders and business groups to garner feedback and exhibitions were held at the community and school levels. It was important for the public to understand that NEWater was a safe and sustainable source of water for Singapore, having been vetted by more than 65,000 water quality tests involving more than 290 physical, chemical and microbiological parameters. This was all the more credible when it came from the experts who had the necessary expertise, and this was fully exemplified through news reports that followed ('NEWater of higher quality than most drinking water in the world – expert', Channel NewsAsia, 16 July 2002, 'Experts endorse NEWater as safe for drinking', Channel NewsAsia, 11 July 2002, 'NEWater gets panel's clearance, Business Times 12 July 2002).

A panel of international and local water experts in the fields of engineering, biomedical sciences, chemistry and water technology provided independent expert testimony and addressed health, safety and quality issues. Their verdict: that NEWater was consistently of a safe, high quality, well within the drinking water requirements of the World Health Organisation and United States Environmental Protection Agency. As experts in their own areas of research who were also familiar with water recycling projects elsewhere, they were best suited to address safety and health concerns, which were the public's top priority.

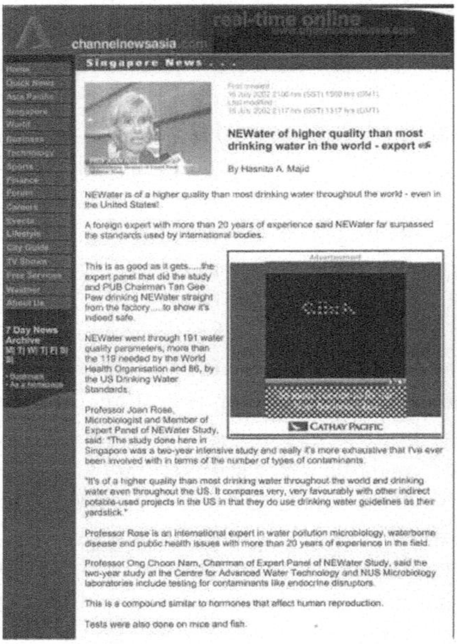

channelnewsasia
Singapore News . . .

NEWater of higher quality than most drinking water in the world - expert

By Hasnita A. Majid

NEWater is of a higher quality than most drinking water throughout the world - even in the United States!

A foreign expert with more than 20 years of experience said NEWater far surpassed the standards used by international bodies.

This is as good as it gets.....the expert panel that did the study and PUB Chairman Tan Gee Paw drinking NEWater straight from the factory.....to show it's indeed safe.

NEWater went through 191 water quality parameters, more than the 119 needed by the World Health Organisation and 86, by the US Drinking Water Standards.

Professor Joan Rose, Microbiologist and Member of Expert Panel of NEWater Study, said: "The study done here in Singapore was a two-year intensive study and really it's more exhaustive that I've ever been involved with in terms of the number of types of contaminants.

"It's of a higher quality than most drinking water throughout the world and drinking water even throughout the US. It compares very, very favourably with other indirect potable-used projects in the US in that they do use drinking water guidelines as their yardstick."

Professor Rose is an international expert in water pollution microbiology, waterborne disease and public health issues with more than 20 years of experience in the field.

Professor Ong Choon Nam, Chairman of Expert Panel of NEWater Study, said the two-year study at the Centre for Advanced Water Technology and NUS Microbiology laboratories include testing for contaminants like endocrine disruptors.

This is a compound similar to hormones that affect human reproduction.

Tests were also done on mice and fish.

NEWater gets panel's clearance

2-year study clears flow of reclaimed water into reservoirs

Central to the entire public education campaign was the setting up of the NEWater Visitor Centre, a state-of-the-art water museum that would act as a one-stop centre for anyone looking to understand how NEWater is produced and how it plays a part in Singapore's water strategy.

This helped visitors to understand the introduction of NEWater in the context of Singapore's water story. Great pain was taken to ensure that the information presented at the center is palatable and easily understood by the public. Technical information on the technology was broken down into layman terms using multimedia animations and interactive touch-screens. For instance, to enable the younger generation to fully understand the process of reverse osmosis, familiar objects like ping pong and tennis balls were used to help illustrate how the process works. To enhance the learning experience for visitors, the NEWater plant was incorporated as part of the centre so visitors could actually see the membranes used in the purification process.

PUB also bottled NEWater in attractive packaging so that the public could sample for themselves how pure the water was, and bottled NEWater was distributed at grassroots and national events, and at the NEWater Visitor Centre. Top government officials, including the then Prime Minister of Singapore, became NEWater ambassadors and champions when they were seen publicly drinking NEWater. This was a huge boost to the campaign, as it positioned the project as a national effort led by the country's leaders. During the 2002 National Day, some 60,000 Singaporeans toasted to NEWater in a show of solidarity, demonstrating the support and confidence they had in NEWater. This was the definitive moment that set the stage for the implementation of NEWater in Singapore. An independent survey by Forbes Research at the end of 2002 confirmed the success of the programme. NEWater had garnered a 98 per cent acceptance rate, with 82 per cent of respondents indicating that they would drink NEWater directly and another 16 per cent would drink it mixed with reservoir water.

Confidence building is not a one-off process and instead of resting on its laurels, PUB continues to sustain its public education campaign. This is critical to avoid being subjected to the 'out of sight, out of mind' mentality by the public who may forget why and how they came to accept NEWater. The NEWater Visitor Centre continues to be the main vehicle for sustained public education on NEWater, with nearly 700,000 visitors to date. In 2005, it clinched the highly recommended award under the International Water Association (IWA) – Marketing and Communication Award for School Information Programme, and the Best Sightseeing/Leisure/Educational Programme at Singapore's 20th Tourism Awards 2005. Key milestones such as the opening of the fourth NEWater plant, commemoration of the 1st and 5th millionth NEWater bottles were publicized to keep NEWater in the public eye and more importantly, maintain the public's confidence in NEWater.

10.6 FROM ZERO TO HERO

From a technology standpoint, PUB's NEWater programme was not considered a major breakthrough. Membrane technology, the heart of the NEWater treatment process has been in use by water managers around the world for decades. It was only in the 1990s that the technology improved considerably in terms of performance and cost, and thus made it viable for PUB to harness the technology for water recycling.

Yet, by building the water reuse campaign on public confidence and acceptance, NEWater has gone from zero to being the hero of Singapore's water sustainability, as well as the global water fraternity. With this success, the NEWater programme sets PUB apart as the pioneer in large-scale water recycling and more importantly, as the international beacon for public education on reclaimed water.

One of the key factors leading to its success was the groundswell of support PUB garnered in the early stages of the campaign. Instead of letting the media run amok with images that would turn the people off, one of the first things that PUB did was to conscientiously engage the media to be on its side. Instead of the 'toilet to tap' imagery, the result was fair and unbiased reporting ('Good as new', TODAY, 11 July 2002). Media reports also helped to frame NEWater in the right context of Singapore's water strategy, ('Four big taps will keep water flowing', The Straits Times, 23 May 2002). Public endorsement by the country's leaders was a powerful and influential signal that served to sway the people in the right direction, and this was duly conveyed by the media ('NEWater proves to be a smash hit with PM Goh', The New Paper, 1–2 August 2002).

It is worthwhile to note here that while the campaign worked out well for Singapore, timing was also of the essence. The introduction of NEWater was timed with the public show of support from the Ministers, Members of Parliament and grassroots leaders which helped to rally the Singaporeans towards its acceptance. This was all the more significant when set against the backdrop of the 2002 National Day celebrations. This show of confidence invokes a sense of nationalistic pride and positions the project as a national effort which needed the support of all Singaporeans. Letters from Singaporeans to the media were in full support of NEWater ('Cheers! Let's drink to NEWater', The Straits Times, 26 July 2002). Future projects looking to emulate the success of PUB will do well to take into account the timing of the campaign and formulate a strategy that would allow them to work it to their advantage.

Since its introduction, NEWater has grown from strength to strength and today it is favoured by the industrial and commercial sectors for its ultra-clean quality. The demand has grown four-fold from 4 mgd in 2003 to some 60 mgd today. By 2060, PUB plans to triple the current NEWater capacity so that NEWater can meet 50% of our future water demand.

REFERENCES

Water Environment and Technology, "Public opposition sidelines Indirect Potable Reuse projects", May 1999.

Macpherson, Linda and Slovic, Paul, "Stigma and Fear – Changing mental models about reuse", 2007.

Chapter 11

Singapore's marina barrage and reservoir – changing mindsets in urban solutions

Brendan Harley
CDM, Singapore

ABSTRACT

... A multi-purpose urban reservoir protects against flooding, boosts water supply, and enhances downtown area ...

Twenty years in development, Singapore's Marina Barrage today is providing three main benefits: as a tidal barrier and pumping station for flood control, as a reservoir for augmented water supply serving some 4.8 million people, and as a new lifestyle attraction around the reservoir and tributary rivers in the heart of urban Singapore.

Despite the challenges – complex system simulation and engineering to assure both water control and water quality, the sheer scale of the barrage crest gates and pumps, and careful construction in one of the world's busiest port areas – Marina Barrage is performing as it was intended to do. It offers a model of integrated and sustainable water management in one of the world's great urban centres.

Keywords: Integrated water management; urban flood control; urban catchments; sustainability; tidal barrage

11.1 ARTICLE

The island nation of Singapore receives some 245 cm of rainfall a year. Monsoon rains can be torrential and sustained – at times creating the risk of severe flooding in downtown areas. And yet ironically, as its limited territory (704 km^2)

constrains the opportunity for retaining natural undeveloped catchment area, Singapore has had to strive to ensure reliable water supplies for its 4.8 million people through innovative water management and investment.

The new Marina Barrage greatly reduces stormwater and tidal flooding of Singapore's vibrant downtown area.

Where before there were tidal flats and mud-banks in old river estuaries, now there is the 240 ha Marina Reservoir, a major new source of freshwater supply. This new reservoir has a capacity of about 10 million m^3 and, when operated for water supply, can support a safe yield of 136,000 m^3/day.

And Marina Barrage creates, at the mouth of the combined estuaries of the Singapore and Kallang Rivers, an urban water feature that stabilizes water levels in the upstream waterways, significantly enhancing urban development and shoreline recreational activities. Some $7 billion of development is presently underway in the Marina Bay neighborhood – all enhanced by the ability of the Barrage to provide stable water levels in Marina Bay.

Since its formal dedication in October 2008, Marina Barrage has played a vital role in helping Singapore to realize the visionary water management and sustainability goals of its leaders.

11.2 20 YEARS IN THE MAKING

For years, low-lying parts of Singapore – including the Little India and Chinatown districts – had experienced flooding during high tide conditions. The flooding could not be avoided other than by rebuilding these areas of Singapore with higher platform levels – or by providing a barrier against the tidal flooding.

PUB, Singapore's national water agency, undertook work on the Marina Barrage project in 1985, as part of a comprehensive drainage management plan that both identified the need for the project as a flood control measure and drew attention to the potential economic and social benefits of stabilizing the water levels in Marina Bay and the Singapore River.

It should be noted that the area surrounding Marina Bay was almost entirely reclaimed land, with lined edges and little long-term ecological development. The tributary canals and rivers, similarly, were lined, urban waterways that had been cleaned up from their earlier conditions through an aggressive PUB campaign.

In the 1990s, PUB conducted additional studies of Marina Barrage and Reservoir, which included evaluating the project, with its heavily urbanized catchment, for its water supply potential. The continued rapid development of Singapore – with a focus around Marina Bay – led to the decision to implement the Marina Barrage project in 2001, and PUB undertook conceptual and final design and construction (Figures 11.1 and 11.2).

Figure 11.1 Aerial view showing Marina Barrage and its iconic pumping station/visitor centre

Figure 11.2 Marina Barrage creates a unique and active public interface with Singapore's water infrastructure – and its vision of integrated water management

The Barrage complex includes the following features:

- A tidal barrage spanning the 350m-wide Marina Channel, comrpising nine 30 m-long hydraulically operated steel crest gates, each gate weighing 70 tonnes

- Seven drainage pumps (six main pumps and one standby), each 7.5 m tall and featuring 3.2 m-diameter impellers turning at 120 rpm – capable of pumping 227 m^3 per second with one pump in reserve
- An electrical building receiving utility power, augmented by three 4.3 MW engine generators capable of powering all loads in the event of a loss of utility power
- A boat hoist at the east terminus of the barrage, to allow transit of small craft (up to 70 tonnes weight) across the barrage
- A 3 m-diameter intake extending some 200 m into the reservoir conveying water to a 5 m^3/sec pumping station that transfers raw water via a 14 km pipeline to Upper Peirce reservoir
- A visitor center presenting 'Sustainable Singapore' in six galleries totaling 2,000 m^2 of interactive display space

The barrage and drainage pumps were sized to allow for moderate effects from climate change, in terms of increasing sea levels over the next 30 to 50 years. The drainage pumps can also handle increased intensities of storm runoff, using the storage capacity in the reservoir to provide some mitigation as well.

11.3 AN ADVANCE IN MULTI-PURPOSE URBAN WATER MANAGEMENT

Marina Barrage represents a unique application of technologies working in combination to achieve results of multiple benefits to Singapore's 4.8 million people. While the use of tidal barriers is well established in locations like Boston (creating the Charles River Basin), London (the Thames Barrier), and the Netherlands (Delta Works), the Marina Barrage operates differently from these in order to fulfill its multipurpose mission in the setting of Singapore's climatological, hydrological and tidal conditions.

In normal conditions, the steel crest gates remain closed in order to isolate the reservoir from the ocean; outflow is directed through low-level sluice gates underlying the pumping station (Figure 11.3).

During heavy rains, and when the tide is low, the gates are lowered sequentially to release excess flows from the stormwater runoff arriving at Marina Basin via the five river systems of its 10,000 ha catchment. When the tide levels are high, however, the gates are raised again and the pumping station is put into operation, capable of moving almost 3.8 million cubic metres per day per pump.

Thus, in response to a range of storm flows and tide levels, the barrage and pump station work in tandem to maintain near-constant levels in Marina Reservoir and at the same time allow natural flushing of salt water as the formerly saline basin is transformed into a fresh-water reservoir. Based on Singapore's past experience in converting coastal estuaries into water supply reservoirs, it is expected that Marina Reservoir will adapt, establishing a new freshwater ecology.

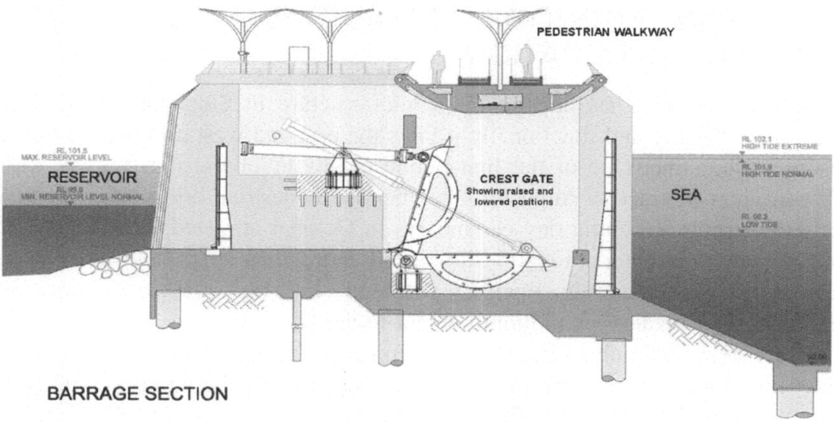

Figure 11.3 Nine 5 m-high crest gates are hydraulically operated, each weighing 70 tonnes

11.4 A MODEL FOR URBAN CENTRES

While unique in its concept, creating a multi-purpose reservoir in the heart of the city, Marina Barrage offers an attractive model for other low-lying coastal cities facing issues of flooding and demand for new water supply. PUB have observed interest in the concept from urban centers like Hong Kong, Dublin, and New Orleans.

The thinking behind Marina Barrage and Reservoir holds value as well. Creating the Marina Reservoir is an important step forward in Singapore's 'Four National Taps' water supply program. In striving to achieve water independence – and facing severe limitations in local sources of supply – Singapore has adopted a strategic initiative that aims to ensure a diversified and sustainable water supply through local catchment water, imported water, reclaimed water (branded as NEWater), and desalinated seawater. Marina Barrage enables a major expansion of supply from local catchments – the first of the 'Four National Taps' (and more cost-effective than desalination or NEWater production) – and will ultimately yield at least 10% of the national supply. This has obvious and near-term benefit to the Singapore public.

But the concept and implementation have great future value to the water management profession as well. As our metropolitan centers face increasing demands for expanded water supply – often, even as impacts of climate variation or climate change reduce yield from traditional sources of supply – the need for integrated water management solutions becomes more acute. And bold, multi-purpose approaches like that represented in Marina Barrage and Reservoir offer tremendous value as models of just what can be achieved. Crisis becomes opportunity.

11.5 SOCIAL, ECONOMIC AND SUSTAINABLE DESIGN CONSIDERATIONS

Already the Barrage has become an iconic structure in Singapore's downtown landscape. Visually and functionally it links the east and west arms of the Marina Reservoir shoreline. One of the first public events involving the Barrage was a half-marathon footrace in August 2008, when the pedestrian bridge served as part of the race route. Opening day ceremonies in October attracted more than 2,500 Singaporeans and international visitors, and more than 40,000 visitors passed through the facility on its opening weekend. In May 2010, PUB officially celebrated the arrival of its millionth visitor to the Barrage.

In fact, everything about the Barrage is meant to attract and engage the public. Water taxis deliver visitors from downtown locations. Public artwork is placed at key points throughout the site. Pulsing fountains and other water play features are open to people of all ages (Figure 11.4). School children's class field trips routinely visit the barrage and learn about its operation. A grass roof provides visual pleasure and valuable open space at the southern terminus of newly developed tropical botanical parklands, Gardens by the Bay.

Figure 11.4 The water theme is beautifully expressed in courtyard fountains and public sculpture throughout the facility's grounds

The visitor center holds six galleries of interactive displays, videos and working models relating the history of the reclamation of the once-polluted Singapore River, demonstrating the development and operation of Marina Barrage, and featuring exhibits promoting the theme of 'Sustainable Singapore' (Figure 11.5).

The 'green roof' holds another feature of the facility's sustainable design – a 70 kW peak DC solar park that provides supplemental power for lighting and general purposes at the visitor center, SCADA control room and common area. It was the first large solar power park of its type in Singapore.

Figure 11.5 Visitors to the Barrage experience 'Sustainable Singapore' in six galleries totaling 2,000 m^2 of interactive display space

Taken together, the concentric helical turf covered roof ramps, the 360-degree views of city and sea, and the attraction of the reservoir itself for water sports or water taxi rides – all serve to advance one of Singapore's strategic goals – to bring residents closer to water, encouraging protection and appreciation of the nation's water resources. In this way, Marina Barrage echoes and reinforces PUB's program for Active Beautiful Clean Waters (ABC Waters) and all of its attendant social benefits.

The project provides a compelling illustration of what can be achieved when an agency like PUB reaches out to the public and invites them to share in the use of a major facility.

11.6 ENGINEERING COMPLEXITY

PUB and its contractors overcame major challenges at every project stage.

In the study stage, complex hydrological and water quality modeling (using SWMM, WASP, and MB-BASIN – a specialized tool developed for this project) was undertaken in order to provide the essential understanding of the hydrodynamics of the Marina Basin and its response to impoundment. Modeling provided the basis for sizing of crest gates and pumps for flood control, for safe yield projections for water supply, and for definition of operational protocols.

The major construction challenge was the installation of the permanent barrage structure within Marina Channel, on a site with two thick layers of marine clay.

As the Channel was to be kept navigable through much of the construction period, the barrage was built in two stages, the first extending west from the East Abutment. Double-skin sheet pile walls, driven to 43 m depth and sand-filled, were adopted to form marine cofferdams within which the permanent structures could be constructed. The barrage structure and pumping station were constructed on a foundation of some 850 1- to 1.5-m diameter bored piles up to 73 m in depth – placed end-to-end, the piles would total almost 65 km in length.

The project team had to contend simultaneously with large-scale construction and precision tolerances. In order to install the massive components of the barrage – the 70-tonne gates and 500-tonne precast bridge segments – floating cranes with lifting capacities up to 3200 tonnes were used. Precision engineering was necessary to achieve tolerance levels of +2 mm for the centerline of the nine pivot pins of each crest gate.

The seven drainage pumps are among the largest in the world. Extensive modeling was undertaken in design of these massive pumps.

As the barrage was entering its final stages of construction, the Marina Channel was no longer navigable to the sea. Barrage construction was sequenced with the ongoing changes in the port usage such that the majority of the marine support traffic was moved to a new pier facility constructed outside the Barrage, and alternative jetty and maintenance facilities also developed elsewhere. The boat hoist at the Barrage was completed in time to ensure uninterrupted egress/access for small work boats and coastguard craft.

The Marina Barrage project, including the visitor centre, was completed at a cost of $250 million.

11.7 A MODEL FOR 'CITIES OF THE FUTURE'

Marina Barrage marks a milestone in PUB's determined quest to secure Singapore's water future as well as to extract maximum advantage from the city-state's water resources – the centerpiece of Singapore's approach to managing water resources in an integrated, sustainable way.

In a time of heightened attention to the life-giving and -enhancing value of our natural resources, innovative, multi-purpose approaches like that represented in Marina Barrage and Reservoir offer a valuable pointer toward intelligent water management in 'cities of the future'.

Part III

Driving Better Design

Chapter 12

Strategic planning for sustainable and integrated urban water management in some SWITCH demonstration cities

Peter van der Steen

UNESCO-IHE Institute for Water Education, Delft, The Netherlands

ABSTRACT

Increasing global change pressures require urban water managers to improve the overall sustainability of their urban water systems. This paper illustrates such an improvement process based on SWITCH demonstration cities Accra, Alexandria and Tel Aviv. A multi-stakeholder platform (Learning Alliance) consisting of a cross-disciplinary team of academic experts and practitioners in these cities went through a process of strategic planning and sustainability assessment. Case studies using total water cycle modelling, quantitative microbial risk assessment and life cycle analysis are described and evaluated to test the hypothesis: '*Design and management of the urban water system based on an analysis and optimisation of the entire urban water system (infrastructure and human organisations, water supply, sanitation, stormwater etc.) will lead to more sustainable solutions than optimisation of separate elements of the system*'. It was concluded that the applied assessment tools are instrumental in widening the boundaries of the system, in including all elements of sustainability and as such they may help planners and decision makers to put investments in the right place to generate maximum benefits.

Keywords: Integrated urban water management; Strategic planning; Sustainability assessment

12.1 INTRODUCTION

Increasing global change pressures, escalating costs and other risks inherent to conventional urban water management are causing cities to face ever increasing difficulties in efficiently managing scarcer and less reliable water resources.

In order to meet these challenges SWITCH (Sustainable Water Management Improves Tomorrow's Cities Health) is facilitating a paradigm shift in urban water management. SWITCH is an EU-funded action research program being implemented and co-funded by a cross-disciplinary team of 33 partners from across the globe, including 17 from the EU and 12 from developing countries. The 'consortium' comprises the academic and urban planning fields, water utilities and consultants. This network of researchers and practitioners are working directly with stakeholders in cities in Africa, Asia, Europe and South-America. The overall goal behind this global consortium is to catalyse change towards more sustainable urban water management in the 'City of the Future'. Demonstrating research and sharing knowledge across a range of different geographical, climatic and socio-cultural settings is expected to lead to global adoption and acceleration of more sustainable solutions.

At the city scale the project is integrating results from academic studies and pilot scale demonstrations of innovative methods for urban water management into new strategic directions or a new strategic plan for the city. The nature of the research and demonstration is such that it covers the entire urban water system (all water bodies, infrastructure, institutions and customers) and the supposition is that this will lead to more holistic strategic plans, based on the state-of-the-art in academic knowledge.

12.2 STRATEGIC PLANNING AND SUSTAINABILITY INDICATORS

12.2.1 The process of strategic planning

Urban planners and managers in cities all over the world are making an effort to steer their city towards a state which we may call 'sustainable'. Though there are many factors within and outside the city outside their control, it is crucial that they have a clear 'Vision' of what a sustainable water system in their city would look like. Moreover, knowledge about the changes that will affect their city in the future and about the strategies that are available to bring the vision to reality, are crucial. For the water system of the city, 'sustainability' has certain characteristics, related to water quality, water availability, energy consumption, health and safety, etc. Plans are made at various levels to undertake actions to achieve a more sustainable urban water system. A Strategic Urban Water Management plan is an overarching plan, with the objective to steer lower level plans such that they all aim in the same direction. A strategic plan also tries to enthuse and to involve public and private parties, commercial and non-commercial into the cities' journey towards sustainability.

Planners and managers face the need to make decisions on new strategies and on new projects. A guiding question for the selection of alternative options should be 'how does it affect sustainability of the system?' To make these decisions, information is required. This information could be in the form of:

- An assessment of the current state of the cities' water system; based on the scoring of a set of agreed 'sustainability indicators'.
- An evaluation of the alternative options in terms of their effect on the sustainability indicators.

Cities that do have a strategic plan monitor the city's advance towards achieving the city's vision by scoring a set of indicators and by publishing the results. Indicators are therefore an integral part of the strategic planning process. A good example in this respect is London, where the Sustainable Development Commission has published reports on 'quality of life' indicators three times since the Sustainable Development Framework, including vision and objectives, was formulated in 2002 (LSDC, 2009). Figure 12.1 shows the strategic planning process undertaken in SWITCH demonstration cities, and how it relates to sustainability indicators.

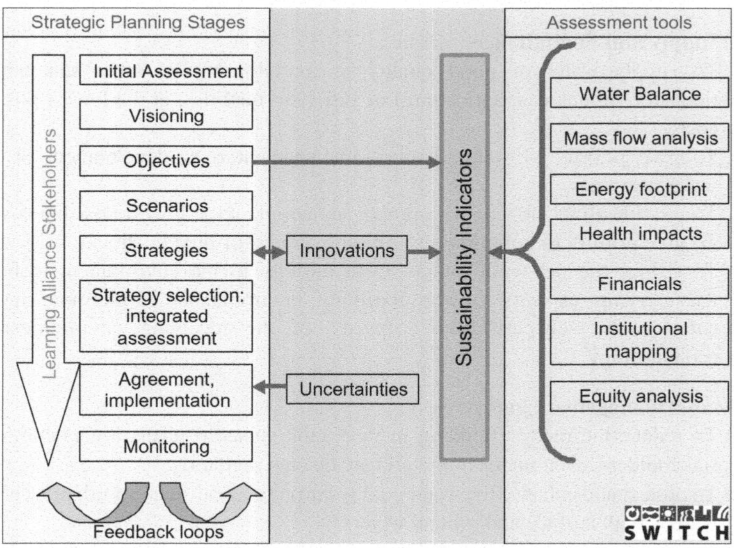

Figure 12.1 Overview of strategic planning exercise in SWITCH demonstration cities

12.2.2 Assessing sustainability

In SWITCH the use of sustainability indicators is directly linked to a Strategic Planning process in a number of SWITCH Demonstration Cities. Based on the experiences in these cities, a general vision for a sustainable urban water system was drafted (Box 12.1). The vision was translated into objectives and indicators

Box 12.1 SWITCH vision of a sustainable urban water system

The SWITCH vision for 2035 for a truly sustainable urban water system was formulated in the form of a set of sustainability objectives. A city that wishes to achieve sustainability for its urban water system should set itself the following objectives:

General

1. To have citizens that are aware of 'water and sustainability' and where the public is involved in decision making.
2. To manage its urban water system in an integrated way; integrating aspects of water supply, stormwater management, wastewater collection, wastewater treatment and wastewater reuse.
3. To use a set of sustainability indicators for decision making, strategy development and planning.
4. To have a strong scientific basis for decision making concerning the management of its Urban Water System. To ensure equity in the access to water, as well as to irrigated green areas.
5. To minimise the energy consumption in the urban water system.

Water Supply and Sanitation

6. To supply water of good quality to its citizens and other stakeholders (industry, companies, agriculture) in sufficient qualities, at the lowest possible costs.
7. To give priority to water demand management over development of new water resources.
8. To provide all its citizens with proper sanitation, at the lowest possible costs.
9. To give priority to pollution prevention over end-of-pipe treatment.
10. To reduce the net waste output from the city to the environment to below the carrying capacity of the receiving environment. Furthermore it will enhance the self-purification capacity of the receiving environment by ecohydrology.

Stormwater management and reuse

11. To reduce the risk of flooding in vulnerable areas to levels acceptable to all stakeholders, even under future climate change scenarios.
12. To protect and enhance the water quality and ecological status of urban receiving waters, both surface and ground waters.
13. To apply source control techniques to enable stormwater to contribute to the quality of life in the urban environment.
14. To harvest rainwater and stormwater for non-potable reuse purposes.
15. To utilise stormwater to re-establish a balanced natural water cycle (in conjunction with landscape development).

that are felt to be relevant for most cities. It is realised that there is no agreed and practical definition of what 'sustainability' of an urban water system means. The definition is contested and the selection of indicators will also reflect this. A purely scientific selection of indicators will therefore not be possible. An attempt was made to this by the MISTRA project (Hellström et al., 2004). They defined sustainability of a water system, a system that does not produce more pollution than the receiving environment can carry. As long as the carrying capacity is not surpassed, the system is sustainable. Though this approach may lead to the definition of some useful environmental indicators, it is not able to define indicators in the field of socio-economics. There are also practical limitations with regard to the number of indicators a city is able to maintain. The set of indicators is therefore always a negotiated set of indicators.

Based on 5 years of experience in the SWITCH cities, the approach shown in Figure 12.1 is advocated, though each city would need to adjust the approach to take into account local conditions. The process consists of two interlinked activities. Firstly, the strategic planning actions carried out by the cities Learning Alliance (LA) or a core-group of the LA. A Learning Alliance is a multi-stakeholder platform for action-research and up-scaling of innovations (Butterworth et al., 2008), and as such is suitable to carry out a strategic planning process. Secondly, the application of assessment tools by the LA or a sub-group of the LA. Academic support is required to apply the assessment tools.

The sustainability indicators are linked to the strategic planning stages as well as to the assessment tools. The vision and objectives determine which indicators are relevant to evaluate strategies. The assessment tools help to score the indicators for the various scenarios and strategies. Innovations (technical, socio-economic or institutional) are also evaluated against the set of indicators. Uncertainties in the score of indicators are presented to decision makers in order to arrive at well-informed and balanced decisions.

A number of assessments have proved useful – examples of those developed and used in SWITCH are provided below:

City water balance. This is the very first information that is required to do any work on the cities' water system. It deals with the current water fluxes (water supply inputs, precipitation, wastewater and stormwater outputs) and the elements of the system (water sources, treatment plants and distribution networks, sewers and drainage, pumping stations etc.) (Last, 2008).

Mass flow analysis. This is the identification and quantification of constituents flowing through the water system, and can be used to make a contaminant balance. It includes organic matter and nutrients, but also toxics like heavy metals or even endocrine disruptors etc. The result of the analysis shows the sources of pollution in the city and quantifies the pollution loads to receiving waters. Mass flow analysis is part of the initial assessment and also part of the evaluation of strategies.

Energy footprint. Consumption of energy in the urban water system has received considerable attention in the last few years. Although it is not one of the larger energy consumers in the city, it still make sense for the water sector to contribute to energy saving. The energy footprint of the urban water system is caused directly by pumping stations, water and wastewater treatment plants and indirectly by in-house heating of water and other consumer activities. In SWITCH a model was developed, based on the energy consumption per unit water, for each of the flows in the water balance. Based on the water balance one can calculate the total energy footprint for the current system, as well as for future scenarios and strategies (Last, 2008).

Health impacts. The major social goal of the water system is to provide health and safety. It therefore makes sense to assess the microbiological and chemical risks of the water system that affect the citizens of the city. Microbial and Chemical Risk Assessment are tools that can be applied to assess the current risk, evaluate the future risks and to assess which interventions are most cost-effective from a health point-of-view.

Finances. The financial situation regarding the urban water system and its users can be captured in a number of financial indicators. Studies on this are ongoing, especially evaluating the effects of new technological options on the financial flows in the system, i.e. who pays for the services and who receives the revenues.

Institutional mapping. The institutional set-up and governance practices in a city are given at the start of a Strategic Planning process but may change or need to change during its implementation. Some new strategies may only be feasible if certain governance and institutional structures are in place.

Social inclusion. Access to water supply, sanitation and a safe living environment is often unequally distributed over the city. At the start of a strategic planning process this should be assessed and monitored during its implementation. Decisions on alternatives should include a social inclusion indicator. For example, Tel Aviv included 'affirmative actions' for the less affluent parts of the city and is in the process of formulating an indicator to measure this (Hoffman et al., 2010).

12.3 SCIENTIFIC ASSESSMENT TOOLS AND SYSTEM BOUNDARIES: EFFECT ON RECOMMENDATIONS FOR URBAN WATER MANAGEMENT

12.3.1 General

In the SWITCH cities research has been carried out using some of the assessment tools shown in Figure 12.1. Starting point for the research has been the application of the tools to a system with boundaries as wide as possible, in line with the general hypothesis for the project: '*Design and management of the urban water system based on an analysis and optimisation of the entire urban water system (infrastructure and human organisations, water supply, sanitation, stormwater etc.) will lead to more sustainable solutions than optimisation of separate elements of the system'.*

In case studies in Tel Aviv, Accra, and Alexandria, the following assessments were applied:

- Application of a Total Urban Water Cycle Model to Develop Urban Water Management Strategies for Tel Aviv, Israel, Water Balance and Energy Consequences. (Tong et al., in prep)
- Quantitative Microbial Risk Analysis (QMRA) to evaluate health effects of interventions in the urban water system of Accra, Ghana (Labite et al., 2010)
- A Life Cycle Analysis of the Urban Water System of Alexandria City, Egypt (Mahgoub et al., 2010).

12.3.2 Water-balance and energy studies for the development of urban water management strategies

Holistic approaches to urban water system management can only be successfully developed if the water balance for the city is known. The water balance consists of incoming water fluxes (precipitation and imported water through the water distribution network) and outputs (evaporation, infiltration, stormwater, wastewater) (Last, 2008) Can this type of basic information help planners and decision makers to decide about strategies for urban water management, aimed at improving sustainability? This question was addressed in research in Tel Aviv, Israel (Tong et al., in prep). The current water balance was developed and the effect of future scenarios (population growth, changes in precipitation, temperature and change in permeability of the city through further urbanisation) and strategies were investigated.

Strategies included rainwater harvesting, stormwater use, permeable pavements and wastewater reuse. The effect of the strategies on total water imported into the city was a reduction of 10% by rainwater harvesting and 32% by wastewater reuse at cluster scale. The latter strategy reduced energy consumption by about 20%, from 2.89 kWh per m^3 of volume of water used (import + reused) in the current situation to 2.45 kWh per m^3.

Does this type of research that takes the entire water system of the city as system, provide information to planners and decision makers that would not be generated in research that addresses a component of the system? The water balance has highlighted the importance of stormwater in the balance, even for a relatively dry climate. The stormwater could in theory satisfy about 50% of the city's water demand. The study showed also how using local sources of water (rainwater, stormwater, treated wastewater) would reduce the need for importing fresh water. These observations are less likely to be taken seriously without an analysis of the entire urban water system. In a case in New South Wales, Australia, this integrated urban planning exercise identified potential cost savings of up to 50% of the capital costs of the system upgrade (Anderson and Iyaduri, 2003).

With regard to the energy costs, the integrated study by Tong et al. showed that energy savings in one element of the system may be offset by increased energy costs in other elements. For instance, the application of on-site wastewater treatment systems was found to increase the overall energy consumption in the system. Other interventions, such as rainwater harvesting or wastewater treatment at intermediate scale had a more positive effect on the overall energy consumption of the system. Analysing water and energy in the entire system, prevents wrong decisions on promoting solutions that may save water but that may increase overall energy spending and thus negatively affect sustainability.

12.3.3 Application of QMRA to the entire urban water system

Contrary to most QMRA studies which are applied either to the water supply system or the sanitation system (Howard et al., 2006; Schönning et al., 2007), the study by Labite et al., (2010) provides the health risks stemming from the entire urban water system in Accra (Ghana). The disease transfer pathways included drinking water supply, contaminated surface waters and open roadside drains. The analysis showed that 88% of the disease burden in Accra (expressed as Disability Adjusted Life Years – DALYs) was caused by shortcomings on the sanitation side of the urban water system, while only 12% was caused by shortcomings in the water supply system.

The effect of a number of interventions aimed at reducing the disease burden was assessed (see Figure 12.2) and it was found that interventions at the sanitation side of the system are more effective than interventions at the water supply system. The efficiency in terms of US$ per DALY averted, of the following three interventions of the Accra Sewerage Improvement project could be quantified (Intervention A: sewerage network and sanitation facilities; Intervention B: sewage network and sanitation facilities combined with treatment plant; Intervention C: coverage of the roadside drains). Intervention C had the highest efficiency. Further work (Machdar et al., in prep) showed that the efficiency of intervention D, further improvement of the water supply system, is significantly lower than interventions at the sanitation side of the system. Interventions outside the water system, such as hygienic food preparation, were in this case not considered.

The results of these studies seem to confirm the hypothesis concerning the advantages of an integrated or holistic analysis of the entire system. The recommendation one could derive from this work towards the urban water managers (planners, decision makers, policy makers) of Accra is to channel the available funds for investments towards the most efficient interventions. Here is where theory conflicts with practice. Currently, the institutional set-up in Accra is

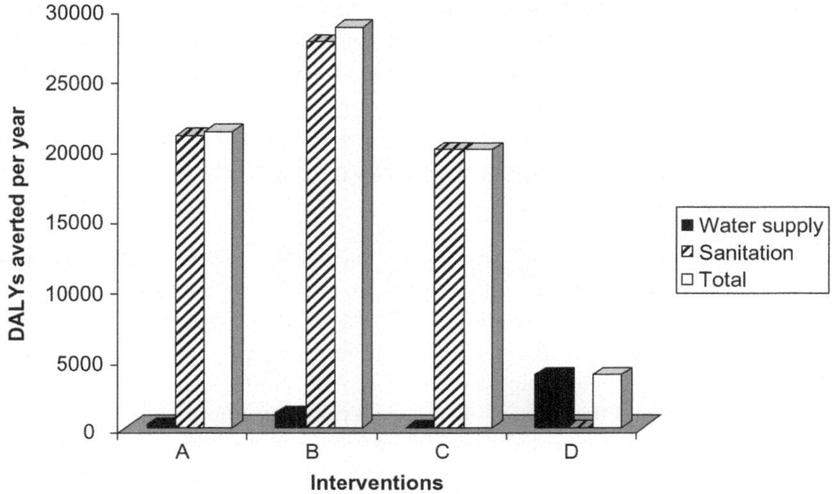

Figure 12.2 DALYs averted after the interventions in Accra's urban water system (A: sewerage network and sanitation facilities; B: sewerage network and sanitation facilities combined with treatment plant; C: coverage of the roadside drains; D: further improvement of the water supply system) (Labite et al., 2010)

that the Ghana Water Company Ltd. (GWCL) jointly with AquaVitensRand Ltd. is responsible for the water supply system. The responsibility for sanitation and the drainage system is with the municipality (Accra Metropolitan Assembly – AMA).

If one would want to implement the recommendation of this scientific assessment tool, the institutional organisation would need to change. The AMA and GWCL would need to coordinate and prioritise the investment of available budgets. It is obvious that this is more easily said than done, for instance because mandates of the organisations are fixed and organisations would not like to transfer budgets to another organisation, even if this would mean that the same budget would have more positive effects for the city's population.

Still, since the budgets are public budgets one would need to make sure that the benefit to citizens is maximised, irrespective of mandates or interests of organisations. The system boundary is the entire city and its inhabitants. An interesting example of such an approach is a QMRA study carried out by Yarra Valley Water in Melbourne, Australia. (Kelly, personal communication). The question was whether investments should be made to provide (more) treatment for the drinking water of the city. To assess whether this was a justified expense of public money a QMRA study was carried out to compare the investment in drinking water treatment with a completely different action, namely spending the same money on a campaign for early detection of breast cancer. The result was that the latter action had a better efficiency in terms of health gain per dollar investment. Rational decision making and rational decision makers would need to take such considerations seriously, even if the system boundaries in this example are extending beyond the water system.

12.3.4 A life cycle analysis (LCA) of the urban water system

While the traditional objective of urban water managers is to provide drinking water of sufficient quality and to prevent pollution of surface or ground water, the LCA methodology includes all environmental impacts of a product or an activity. In that sense the method fits well with a holistic approach to urban water management. The urban water system of Alexandria was modelled using SimaPro software, and the impacts of various options to improve the system were analyzed based on the Eco-indicator 99 assessment method (Mahgoub et al., 2010). The results show that the highest impact in today's system is generated by the disposal of primary treated wastewater (68% of the total impact) and by high-energy consuming water treatment plants (18% of the total impact). Various options for improvement of the water systems, such as leakage reduction, upgrade from primary to secondary wastewater treatment, separation of household wastewater at source and decentralised treatment and sewerage expansion, were investigated. The results show that the reduction in environmental impact of wastewater system upgrading to nutrient removal and effluent reuse has a similar result as changing to a decentralised system of household separation and decentralised treatment of black and grey water. Another interesting result for the discussion here is that the option of sewerage expansion (from about 90% to 100%) would not reduce the total environmental impact, basically because only primary treatment without nutrient removal does not significantly reduce the environmental impact – a limitation of the measurement approach of the Eco-indicator 99 method.

How could the planners and decision makers of the Wastewater and Drainage Company of Alexandria make use of this kind of information? Firstly, it allows them to prioritise investments in those measures that minimise the environmental impact. Secondly, it allows comparison of quite different approaches, such as effluent reuse and effluent disposal. Thirdly, the results also make clear that expanding the sewer system without providing nutrient removal has no positive overall effect of the environment and can therefore only be a temporary situation. It could even be strategic to postpone the expansion of the sewer until sufficient funding has been ensured for implementation of nutrient removal.

A disadvantage of the LCA methodology, is that it does not consider social disparities within the system, i.e. low income areas. In such areas the various benefits of some kind of sewerage or sanitation make perfect sense, even without achieving any overall environmental benefit, as measured by the LCA.

12.4 LESSONS LEARNED IN SWITCH DEMONSTRATION CITIES ON STRATEGIC PLANNING FOR THE URBAN WATER SYSTEM

The demonstration cities that are involved in the SWITCH strategic planning exercise (Accra, Alexandria, Cali, Tel Aviv, Hamburg, Lodz) have each prepared visions for their city, while most have also defined the objectives, sustainability indicators, scenarios and strategies. The lessons learned can be summarised as follows:

- The strategic planning method suggested to the LA's is generally well accepted. The concept of joint strategic planning, involving a cross-disciplinary team of academic experts and practitioners was uncommon to basically all LA's. The strategic planning exercise helped to give focus to the LA process. In many cities it was the first time that a group gathered of mixed composition with representatives of most water sector institutions and academics. Some staff of water sector institutions met each other for the first time. Though the strategic planning exercise was initiated from outside (i.e. from SWITCH), in all cities there is a growing ownership of the process.
- It is clear in all cities that professional support by an interdisciplinary team with expertise in planning processes, institutional change, policy influence and political science (in addition to technical specialists) is required for a successful exercise.
- It is important that the LA members include sufficient numbers of higher management with responsibility for strategic planning in their organizations.
- As long as the strategic planning process is not part of an 'official' process, then it is a flexible process that can take place in various political realities. However, at some point formal commitment to the process is required, in order for the results to be taken up in official policy documents. The cities municipality is usually the key-organisation in this respect.
- Visions were defined without many disagreements, but this could change when the process becomes more official and when strategies are going to be defined.
- Communication between city professionals and academics needs proper facilitation, in both directions and preferable by a neutral party. The inclusion of scientific innovations in strategic plans is not automatic, even though the potential benefit for the city may be obvious for the developers of the innovation.
- The use of sustainability indicators is in general welcomed by the LAs. Sometimes, the use of indicators is simplified too much, i.e. seen as a goal in itself rather than a tool to be used in a strategic planning process.

- Formulating the strategic plan in itself is not sufficient to cause a 'transition' to a sustainable urban water system. A transition strategy about how to best implement innovative methods of urban water management is therefore required (Duffy and Jefferies, in prep).

12.5 CONCLUSIONS

It is suggested that sustainability of urban water systems can be improved by application of a joint process of stakeholder driven strategic planning, together with scientific assessment of the systems sustainability. Proper assessment positions the system borders wide enough, and includes all elements of sustainability, to encourage rational decision making. This may help planners and decision makers to put investments in the right place to generate maximum benefits.

REFERENCES

Anderson J. and R. Iyaduri (2003) Integrated urban water planning; big picture planning is good for the wallet and the environment. Water Science and Technology 47, 7–8, 19–23.

Butterworth J.A., A. Sutherland, N. Manning, B. Darteh, M. Dziegielewska-Geitz, J. Eckart, C. Batchelor, P. Moriarty, T. Schouten, C. Da Silva, J. Verhagen and P.J. Bury (2008) Building more effective partnerships for innovation in urban water management. International Conference on Water and Urban Development Paradigms: Towards an integration of engineering, design and management approaches, 15–19 September 2008, Katholieke Universiteit Leuven, Belgium.

Duffy A. and C. Jefferies (in prep) Developing a Framework to guide Urban Water Systems Transitions. University of Abertay.

Hellström D., M. Hjerpe and D. Van Moeffaert (2004) Indicators to assess ecological sustainability in the Urban Water sector. Chalmers University of Technology, Gothenburg, Sweden, MISTRA Report 2004:3.

Hoffman D., A. Aharoni, D. Jackman, T. Gavrieli, G. Segal, P. van der Steen and A. Adin (2010) Strategic Planning for the Urban Water Systems of Tel Aviv-Yafo Phase II: Vision and Goals and Phase III: Actions, Means and Indicators. Final Report.

Howard G., S. Pedley and S. Tibatemwa 2006 Quantitative microbial risk assessment to estimate health risks attributable to water supply: can the technique be applied in developing countries with limited data? J. Water Health 4(1), 49–65.

Labite H., I. Lunani, N.P. van der Steen, K. Vairavamoorthy, P. Drechsel and P. Lens (2010) Quantitative Microbial Risk Analysis to evaluate health effects of interventions in the urban water system of Accra, Ghana. Journal of Water and Health 8, 3, 417–430.

Last E. Developing a scoping model for sustainable urban water. Biotechnology News 2008, 55:3–4.

LSDC (2009) London's Quality of Life Indicators. 2008–09 Report [http://www.londonsdc.org/documents/qol_reports/QoL_indicators.pdf, accessed November 2010].

Mahgoub M.E.M., N. Peter van der Steen, Khaled Abu-Zeid and Kala Vairavamoorthy. (2010) Towards Sustainability in Urban Water: A Life Cycle Analysis of the

Urban Water System of Alexandria City, Egypt. Journal of Cleaner Production 18, 1100–1106.

Schoöning, C., T. Westrell, T.-H. Stenström, K. Arnbjerg-Nielsen, A.-B. Hasling, L. Høibye, and A. Carlsen 2007 Microbial risk assessment of local handling and use of human faeces. J. Water Health 5(1), 117–128.

Tong T.H.D., N.P. van der Steen, A. Adin and K.Vairavamoorthy (in prep) Application of a Total Urban Water Cycle Model to Develop Urban Water Management Strategies for Tel Aviv, Israel, Water Balance and Energy Consequences.

Van der Steen P. and C. Howe (2009) Managing Water in the City of the Future; Strategic Planning and Science. Reviews in Environmental Science and Bio-Technology, 8, 2, 115–120.

Chapter 13

Water centric cities of the future – towards macro scale assessment of sustainability

Vladimir Novotny[1] *and Eric V. Novotny*[2]

[1]*Northeastern University, Boston, MA, USA*

[2]*Barr Engineering Corporation, Minneapolis, Minnesota, USA*

ABSTRACT

The developments of several current and future ecocities have been comparatively assessed as to their environmental benefits of preservation or restoration of the ecological and hydrological (water reuse) functionality of their surface water bodies and social benefits such as carbon emissions reduction, recreation, and resource recovery. These evaluation categories are, in addition to economic assessment, the key components of the Triple Bottom Line assessment of sustainability. They are based on the urban metabolism concept. The assessed urban water developments include downtown San Antonio (TX, USA), Treasure Island in San Francisco and Sonoma Mountain Village (CA, USA); Hammarby Sjöstad (Sweden); Tianjin, Qingdao and Dongtan (China); and Masdar (UAE). Because of their frugality with respect to energy and water requirements, the 'new cities' can be built in 'hostile' environments such as arid areas with poor soils or decontaminated brownfields. The analysis revealed some problems with the lack of macroscale measures, models, and indices for some key components of the triple bottom line assessment that should be investigated and solved by research such as what is the limit of recycle to prevent accumulation of some harmful pollutants within the system, what are the thresholds, and what is the best distributed/hybrid configuration of the water/stormwater/used water management in the Cities of the Future, considering general and local aspects.

Keywords: Ecocity; Ecoblock; Sustainable development; Urban metabolism; Sustainability footprints; Triple bottom assessment; LEED criteria; One Planet Living criteria; Water reuse; Recycle; Global warming

13.1 INTRODUCTION

The concepts of the new paradigm of sustainable water centric ecocities have been emerging for the last fifteen years in environmental research and landscape design laboratories in several countries, including Europe (Sweden, Germany, United Kingdom, The Netherlands), Asia (Singapore, China, Japan and Korea), Australia, United Arab Emirates (Masdar), USA (Chicago, Portland, Seattle, Philadelphia, San Francisco) and Canada (British Columbia). This paradigm is based on the premise that urban waters are the lifeline of cities and focus of the movement towards more sustainable cities. The evolution of the new paradigm of urbanization ranges from microscale 'green' buildings, subdivisions or 'ecoblock' to macroscale ecocities and ecologically reengineered urban watersheds and incorporating also transportation, food production and consumption and neighborhood urban living (Novotny, 2008). Many concepts developed by landscape architects incorporate surface water bodies as a focus. At the same time, environmental engineers and urban planners are developing water/stormwater/wastewater management concepts based on switching from the linear once through water management (minimum reuse) to a closed loop hydrological cycle system that maximizes reuse and recycling. Reduction of green house gas (GHG) emissions and application of green technologies have become a major goal in the last five years.

Water centric sustainable urban developments recognize the ecological value of surface water resources. In this approach the ecological integrity of the water resources and the riparian and flood zones is preserved or restored, using integrated resource management while also considering the impact on GHG emissions. Sustainability implies intergenerational preservation of natural resource assets so that future generations are not adversely impacted by present (and near past) development and economic utilization of resources; it considers the impacts of population increases, occurring mostly in the cities, and effects of global warming.

An *ecocity* is a city or a part thereof that balances social, economic and environmental factors (triple bottom line) to achieve sustainable development (Figure 13.1). Its definition was coined by Register (1987) as 'a **sustainable city**, or **eco-city** is a city designed with consideration of environmental impact, inhabited by people dedicated to minimisation of required inputs of energy, water and food, and waste output of heat, air pollution – CO_2, methane, and water pollution'. Hence, the water centric ecocity will combine and protect the hydrological and ecological value of the urban landscape with sustainable development. Integrated resources management (IRM) of the water centric cities of the future will consider; (1) Water conservation (green development); (2) Distributed stormwater management using best management practices of rainwater harvesting, infiltration and storage of excess flows, and mostly or fully surface drainage; (3) Distributed wastewater treatment generating water for reuse in buildings, landscape irrigation and ecological flow of existing or restored water bodies; (4) Using landscape and landscape components (e.g. ponds, wetlands, grass filters, etc.) for attenuation of diffuse pollution and post treatment of effluents recovered for reuse; (5) Heat and energy recovery; (6) Nutrient recovery; (7) Biogas recovery and, in the future, hydrogen generation from biogas and wastewater in fuel cells; and (8) Extent of use of alternate renewable energy sources.

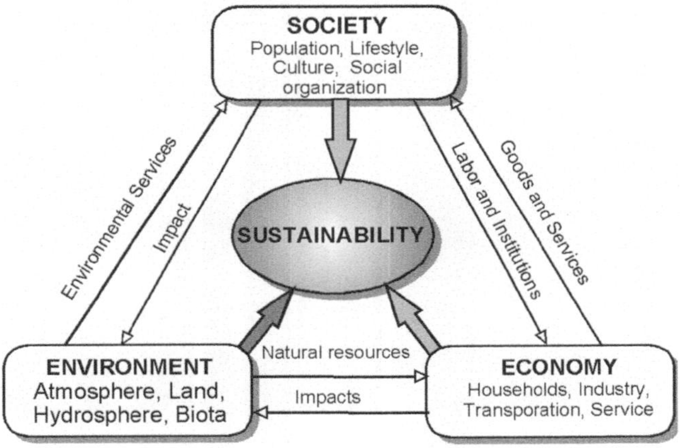

Figure 13.1 Triple bottom assessment and interactions of sustainable developments

Generically, urban planning and its relation to water can range from (A) Cities that develop their water resources to provide visual enjoyment and attraction of surface water bodies with less or even no emphasis of ecological and hydrological functionality of the surface water bodies and less or no reduction of GHG emissions such as, e.g. San Antonio (Texas) or partially Ghent (Belgium), to (B) Sustainable developments, retrofits and ecocities that result in enhancement of the ecological integrity of their receiving surface water bodies, restore the hydrologic function of the watershed, and significantly reduce their carbon footprint. Category (A) developments are mostly spurred by the desire of the cities to bring 'Venice-like' attractiveness to the city, including restaurants, boat rides and gondolas, to attract resident population and tourists. This was facilitated by the cleanup of surface waters mandated by the water quality regulations such as the Clean Water Act in the US and Water Framework Directive of the European Community. Category (B) cities, among other resource restoration and management measures and in addition to aesthetic improvements, recreate ecology of existing or daylighted surface waters currently buried in sewers or culverts and incorporate into their landscape ponds for storage and treatment of urban runoff and/or wetlands used for post treatment of effluents and stormwater.

Category (A) urban developments may not be fully sustainable because the triple bottom line assessment benefits are not balanced, the benefits are mostly economical (tourism, employment in hotels and restaurants), less social (aesthetic, attractiveness of river front development, and noncontact recreation) and not fully environmental (Figure 13.2).

Figure 13.2 San Antonio River (TX) is a beautifully landscaped and restored urban river that brought great economic and social benefits to the city. However, in 2007 the river in the city was a concrete channel with almost no habitat and relatively poor water quality. This type of development does not reduce GHG footprint (Photograph V. Novotny)

13.2 ASSESSMENT OF SUSTAINABITY

13.2.1 Microscale assessment indices and metrics

Several rating systems and performance criteria for assessment of green/sustainable developments have become popular in the last five years. They are a mix of numeric performance metrics with narrative criteria. In general, they focus on individual buildings, limited scale development (for example, a shopping mall or a treatment plant) or a subdivision. The most popular in the USA are:

LEED Criteria. The US Green Building Council has proposed LEED (Leadership in Energy and Environmental Design) standards for 'green' buildings and neighborhoods (USGBC 2005; 2007) that are becoming a yardstick for building and development. The problem with LEED standards is that only about 15% of points obtained in the certification are related to water, water quality and ecology. The LEED standards address:

- *'Green' certification* formulated for homes, neighborhood development and commercial interiors;

- *Smart location & linkage* which include, among others, required indices of proximity to water and wastewater infrastructure, flood plain avoidance, endangered species protection, wetland and water body conservation, and agricultural land conservation;
- *Neighborhood pattern and design* such as compact development, diversity and affordability of housing, walkable streets, transit facilities, access to public spaces, or local food production;
- *Green construction & technology;* and
- *Innovation & design process.*

Low Impact Developments. LID concepts are used in and restricted to subdivisions or small size developments practicing mostly on-site stormwater containment, storage, infiltration and conveyance. In the US, a 'subdivision' is a settlement of tens or hundreds mostly single family homes developed by a single developer who acquires land and installs infrastructure such as roads, drainage, sewers, and, in most cases, water supplies where private wells are not feasible. Some developers also build the homes in the development. The governance of the subdivision is usually by an association the homeowners are required to establish. The goal of LID is to mimic the predevelopment hydrology by minimizing and attenuating urban runoff by implementing green roofs, raingardens, infiltration ponds and wetlands, pervious pavements and other best management practices that control volume and pollution by urban runoff (Prince George's County, 1999). LID low density developments are often situated in rural settings with very high open/built space ratios, which could imply long distance travel and urban sprawl.

13.2.2 Need for macroscale criteria and assessment

The development of the cities of the future, the ecocities, requires a comprehensive and hierarchical macroscale approach to the microscale and often fragmented piecemeal transformation (Hill, 2007) of the current unsustainable urbanization to the new ecology friendly and sustainable urban areas and finally entire cities. The macroscale goals of the new paradigm for water centric ecocity communities are (Novotny, Ahern and Brown, 2010; Novotny and Brown, 2006):

- Developing an urban watershed and its landscape that is sustainable and resilient over a long run and preserves or mimics but not necessarily reproduces the hydrologic processes and ecological structures present in the predevelopment natural system;
- Protection of the natural systems and restoration of natural drainage (daylighting);
- Mimicking predevelopment ecology and hydrology, relying on reduction of imperviousness, increased infiltration, surface storage and use of plants that retain water (e.g. coniferous trees);

- Developing or restoring interconnected green ecotones (green areas bordering the streams that connect nature with the built human habitat), especially those connected to water bodies, that provide habitat to flora and fauna, while providing storage and infiltration of excess flows and buffering pollutant loads from the surrounding inhabited, commercialized, and traffic urban areas;
- Adaptation to the trends of global warming and stresses caused by increasing population. It is not enough to keep emissions at the present level; the new development must dramatically reduce carbon emissions and increase resources to accommodate anticipated urban population increases; and
- Retrofitting and reconnecting old underground systems interlinked with the daylighted or existing surface streams.

The development of macroscale criteria of urban sustainability is a work in progress by several associations, governmental agencies (for example, in Australia) and research groups. A subcommittee of The International Steering Committee for the Cities of the Future of the International Water Association is developing concepts of macroscale sustainability criteria (Novotny, 2010).

Sustainability of the cities, pollution, social qualities and other attributes and amenities are related to 'urban metabolism' (Kennedy, Cuddihi, and Engel-Yan, 2007) which is a mass balance concept that relates inputs into the city (materials, chemicals, water, energy, and food) to the outputs. Outputs are numerous and examples of undesirable environmental include:

- liquid sewage, industrial wastewater and combined sewer overflows (point sources of pollution);
- polluted urban, construction sites and highway runoff (diffuse sources);
- air pollution emissions, including greenhouse gases (GHG);
- rubbish and other solid waste; and
- contaminated land (brownfields) and shallow groundwater.

Urban metabolism can be linear, cyclic or hybrid. Daigger (2009), Novotny (2008) and others agree the current 'linear' approach, sometimes called the *take, make, waste* approach in the sustainability literature, when applied more broadly to natural resources use and global climatic change, has become increasingly unsustainable. If every city on earth would strive to achieve the same consumption of inputs as the North American cities today, the productive land needed to provide food, raw materials, assimilation of waste and emissions, and produce energy would be three times of all available productive land and resources on the earth. Scientists, the informed public and politicians agree that the system must change along the three R's – *reduce (conserve), reclaim, and reuse* which is a foundation of the cyclic/hybrid system.

The footprints for which the sustainability criteria should be developed are based on the Triple Bottom Line (TBL) – Life Cycle Assessment, i.e. they should cover economic, social and environmental/ecological aspects (Figure 13.1) in the intergenerational sustainability context. Currently, the major large scale ('giant') footprints suggested in the literature are (Hoekstra and Chapagain, 2008):

- *Ecological footprint* is a measure of the use of bio-productive space (e.g. hectares of productive land needed to support life of one person in the cities)
- *Carbon footprint* is a measure of the impact that human activities have on the environment in terms of the amount of GHG emissions measured in units of carbon dioxide in tons/person-year
- *Water footprint* measures the total water use on site and also virtual water in liters/person-day.

The One Planet Living (OPL) criteria listed below are an example of large scale criteria of urban sustainability. Introduced by The World Wild Life Fund (WWF, 2008), the criteria promote implementation of principles that include social, environmental and technological TBL metrics as follows:

- net zero carbon emissions with 100% of the energy coming from renewable resources;
- zero solid waste with the diversion of 99% of the solid waste away from landfills;
- sustainable transportation with zero carbon emission from transportation inside of the city;
- local and sustainable materials used throughout the construction;
- sustainable foods with retail outlets providing organic and or fair trade products;
- sustainable water with a 50% reduction in water use from the national average;
- natural habitat and wildlife protection and preservation;
- preservation of local culture and heritage with architecture to integrate local values;
- equity and fair trade with wages and working conditions following the international labor standards; and
- health and happiness with facilities and events for every demographic group.

Some ecocity developments are now aiming at receiving OPL certification. These macroscale criteria for ecocities are more broad and stringent than LEED or LID Criteria. However, there are some equity problems between the developed and undeveloped nations when applying these criteria. For example, the water use criterion is extremely easy to achieve in the US and Canada which have a very high water use (~500 Liters/person-day) while it may be difficult in developing nations where per capita water use may be less than 100 Liters/capita-day or Japan or Europe where typical water use is around 150 Liters/person-day. Also, the OPL criteria do not *a priori* promote urban and suburban agriculture that is important for reusing used water and recovered fertilizer thus reducing the high *virtual* water use to produce imported food for the city.

13.3 SEVEN ECOCITIES CASE STUDY

The authors in Novotny and Novotny (2009) (see also Novotny, Ahern and Brown, 2010) analyzed seven cities/urban developments that are striving to become 'ecocities' and looked at common features of the key macroscale parameters such as population density, energy use and carbon imprint, water use and reuse, and cost. None of the analyzed cities have been fully built. We looked at whether the water system is linear and centralized or closed and decentralized. The results are summarized in Table 13.1.

13.3.1 Hammarby Sjöstad (Sweden)

Hammarby Sjostad, located on Lake Hammarby Sjö in Stockholm, is an ongoing development that shaped the project's infrastructure, planning and design of the buildings into a modern mixed-use sustainable and environmentally friendly urban space. The scheme has attracted international acclaim for the quality of habitat it created and convinced many that carbon neutral development does not require lifestyle changes. The development concept successfully connects the historic landscape with aquatic areas which act as storm water drainage, encourages biodiversity, creation of new habitats, informal amenity areas and formal areas of public open space (Figure 13.3). Sustainability is also enhanced through the use of green roofs, solar panels, and eco-friendly construction products. The use of glass as a core material maximizes sunlight and views of the water and green spaces. The city has a fully integrated underground sanitary (separated) waste collection system conveying wastewater to the local district treatment and heat and nutrient recovery plant. Methane is recovered from digested sludge and organic solid waste is incinerated in a municipal incinerator that recovers energy. Passive energy savings and solar panels, both thermal and photovoltaic are installed throughout the city to save energy. The development has its own ecosystem, known as the Hammarby Model. In spite of the fact that the ecocity is still based on the linear model and water reclamation and water reuse is not included to a great extent, Hammarby Sjöstad is the first city built on ecological principles that broke the barrier toward sustainable urban development.

13.3.2 Dongtan (China)

Dongtan was planned to be at the eastern tip of Chongming Island at the mouth of the Yangtze River in the middle of a designated nature reserve with outstanding

Table 13.1 Synthesis of the main parameters of the seven investigated ecocity developments (Novotny and Novotny, 2009)

City	Population Total	Population Density #/ha	Water use L/cap-day*	% water recycle	Water System	% Energy savings and renewable energy production	Green area m^2/cap	Cost US$/ unit**
Hammarby Sjöstad	30,000	133	100	0	Linear	50	40	200,000
Dongtan	500,000 (80,000)++	160	200	43	Mostly Linear	100	100	~40,000
Qingdao	1500–1800+	430 – 515	160	85	Closed loop	100	~15	?
Tianjin	350,000 (50,000)++	117	160	60	Partially closed	15	15	60,000 – 70,000
Masdar	50,000	135	160	80	Closed loop	100	<10	1 million
Treasure Island	13,500	170	264	25	Mostly Linear	60	75	550,000
Sonoma Mountain V.	5,000	62	185	22	Mostly Linear	100	20	525,000

* Indoor and outdoor domestic water use
** Based on average 2.5 members per household unit
+ Qingdao ecoblock
++ Phase I

biodiversity about 40 km north of downtown Shanghai. The approach to the design of the city by the architectural firm Arup was different from other designs submitted by various other firms in the mode of 'low impact' spread out subdivisions. Arup envisioned Dongtan to be a vibrant city with green 'corridors' of public space ensuring a high quality of life for residents. The city was designed to attract employment locally across all social and economic demographics in the hope that people will choose to live and work there. This pioneering approach initiated the new paradigm of ecocity building that was then adopted by other ecocity developers. The city design is exceedingly water centric with canals and lagoons within the development for aesthetics, recreation and transportation and also water for reuse (Figure 13.4). Energy would be produced by incinerating rice husks and organic solids as well as by renewable solar panels and wind energy production throughout the city.

Figure 13.3 Hammarby Sjöstad (Stockholm) with surface drainage (Photograph credit Malena Karlsson, GlashusEtt, Hammarby Sjöstad). The surface drainage collect all clean water flows (rain, groundwater, cooling)

The development of Dongtan was supposed to have its first 5000 inhabitants moved in during the 2010 World Exhibition in Shanghai but the project realization was indefinitely postponed because of political reasons.

13.3.3 Qingdao (China) Ecoblock and Ecocity

The ecoblock concepts were developed by the team of Professor Harrison Fraker of the University of California Berkeley College of Environmental Design specifically for urban developments in China but could be used anywhere in the world, especially in the countries with rapid population increase. An *ecoblock* is self-sustained and semi-independent with its water and energy needs. It generates its own energy from renewable sources and used water, harvests rainwater, and processes and reclaims its used water.

Figure 13.4 Architect's rendering of the East Village and Lake in Dongtan. Note solar panels on the roofs of the buildings and wind turbines surrounding the lake (Source and courtesy Arup)

A typical standardized ecoblock (Figure 13.5) has 600 units on 3.5 hectares and houses 1500–1800 residents. The proposal for the Qingdao ecocity included 16 ecoblocks. Similarly to Dongtan, the Qingdao ecocity's development has been postponed but the ecoblocks and their concepts have been incorporated into the design of Tianjin.

13.3.4 Tianjin (China)

The site of the new city development is about 150 kilometer southeast of Beijing and 40 km from the historic Tianjin City (population about 12 million) which is the regional center and the largest port city in northeast China. The city will be a part of a huge regional development of the Tianjin – Binhai New Area. The ecocity is a joint venture project of China and Singapore.

The city is divided into (eco) blocks. The smallest block unit has an area of 400 × 400 meters (16 ha). Professor Harrison Fraker (University of California-Berkeley) confirmed that Tianjin would also include the Quindao ecoblocks. The city is water centric and features an 'Eco-valley' which is the main north-south green connector in the city. The Eco-valley will retain a large ecological wetland set aside as a habitat for bird migration, and will preserve former watercourses.

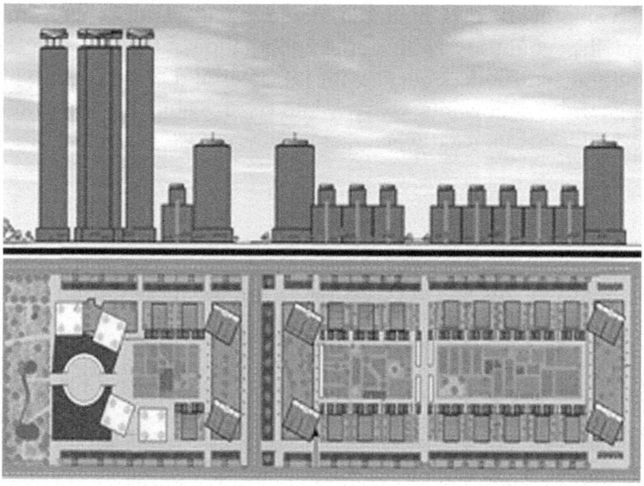

Figure 13.5 Plan and view of the Ecoblock module (Courtesy Prof. H. Fraker, University of California, Berkeley). The treatment wetland is to the left of the ecoblock

The primary sources of water for the new Tianjin ecocity are desalinated water and rainwater which constitute more than 50% of water used in the city. An extensive system of rainfall collection and sewage reuse will be established relying heavily on the landscape for collection and infiltration of rainwater. The city will have centralized sewage and wastewater treatment and recycling and will develop and utilize non-conventional water resources such as recycled storm and rain water and desalinated seawater in a water supply infrastructure that will reduce the need for conventional water resources. Treated used water will be used for landscape irrigation and for the lake and created surface channels.

13.3.5 Masdar (UAE)

Masdar in United Arab Emirates is being built and has been designed to follow the 'One Planet Living' (OPL) ten principles. When the city is completed it will be home to 50,000 residents and 40,000 people are expected to commute to work in the city. Water used inside the city will be provided by a desalinization plant run by solar power which also provides most of energy to the city. The plant will produce two types of high quality water: one fit for drinking and the other fit for personal uses such as showering and washing dishes.

Masdar city will use a plethora of water management principles in order to treat all parts of the water cycle and use them as a water source. As many as nine water conveyance systems will be used in 12 different ways and treated at three treatment levels. The variety of water sources to be used includes groundwater, seawater, surface runoff, rainwater harvesting, dew/fog capture, grey and black water reuse and resource recovery from urine streams. The city will employ extensive passive energy savings by shading, narrow shaded streets, wind exposure for cooling, and solar power plants. Nutrients are recovered and some used to farm algae.

13.3.6 Treasure Island (San Francisco, California, USA)

Treasure Island is a manmade island built by dredging sediments from the San Francisco Bay. Currently, it is a partially abandoned navy base with brownfield problems which are being remedied. The city is embarking on developing the island into a sustainable residential/commercial community with suburban agriculture. The planned water/wastewater system will be linear with some reuse after treatment in a central WWTP for irrigation. 75% of the effluent will be discharged into the bay. Potable water will be imported from the San Francisco municipal grid. Stormwater management will center on xeriscape, permeable surfaces and pavements, green roofs and routing excess runoff to be treated in a wetland. Once the excess runoff is collected it will be routed to a constructed wetland for treatment and water reused for irrigation and other nonpotable uses.

13.3.7 Sonoma Mountain Village

Sonoma Mountain Village in California is in the initial phases of the development. When finished in 2020 it will have 5000 inhabitants, hence, it will be the smallest of the seven ecocities (Figure 13.6). The developer has applied for and received OPL certification. The goal for water used within the village is a reduction in water consumption by 60% from a general norm for single family homes in the region. This will be accomplished through water reduction devises, education, rainwater harvesting and reuse of water. The municipal drinking water supply will be used inside of all buildings and for irrigation in private backyards. Reclaimed water will be used for irrigation of all public parks, medians, and street trees along with high efficiency (sub-drip) irrigation of all common areas, private front yards and for use in firefighting. Stormwater reuse will be used for habitat maintenance, groundwater recharge and as a supplemental irrigation supply for all landscape areas. There will be habitat protected bioswales acting as wetlands connected to an underground reservoir from which water will be recycled for irrigation purposes. Most electricity will be produced by on site photovoltaic power plant and solar panels to achieve net zero GHG emissions goal.

Other 'ecocities' have been proposed and are being developed throughout the world. The first phase of the Dockside Greens in Victoria (British Columbia, Canada) has been essentially completed. It is a hybrid water centric community that has received Platinum LEED certification. It reuses water for providing ecological flow to recreated streams and irrigation and recovers energy from methane from sludge and organic solids digestion and syngas (mostly carbon monoxide) from organic solid waste (O'Riordan et al., 2008). New developments are in the planning stages in Portugal, United Kingdom (2010 Olympic sites), Canada, Turkey and other countries.

Figure 13.6 Architect's rendering of the main square in the Sonoma Mountain Village (Source and courtesy Sonoma Mountain Village – SOMO)

13.4 SYNTHESIS

We have summarized the basic parameters of the analyzed ecocities and the synthesis is presented in Table 1 below.

Population Density. With the exceptions of the Qingdao ecoblock which is the development with the highest population density, the densities of the remaining six developments varied between 60 to 170 people/ha. From the presentations and literature findings it was evident that all design teams used some kind of a proprietary model which balanced the population and its energy use based on probability of walking and biking instead of driving, energy insulation of buildings and exposure to sun, renewable energy sources and other determinants for GHG emissions from urban areas. The fact of medium design density development being the most optimal refutes, to some degree, the utility of the 'low impact' LID subdivisions which in most cases have an objective of minimizing stormwater impacts and discharges and generally results in low density developments but with a relatively high energy use and reliance on automobiles.

Green House Gas Emissions (carbon footprint). Table 1 shows the energy savings and production of energy from local renewable sources (solar, wind) reducing the GHG emissions. For example, Qingdao, Masdar and Sonoma Mountain Village designs are proving ecocities could fulfill the OPL criterion of net zero GHG emissions from infrastructure heating and cooling and electricity consumption. All three developments also reduce energy by restricting traffic inside the city and in Masdar and Tianjin to vehicles powered by electricity produced by renewable sources. Hence, 100% energy savings and renewable energy use implies the net zero OPL criterion is met.

Water Reclamation and Reuse. All cities use the latest technology for in-house water savings such as low flush toilets, showers, etc. Hammarby Sjöstad is almost a 100% linear system with recovery of phosphorus. The Stockholm metropolitan area, where the ecocity is located, is water rich and there is apparently no need for recycle, yet, they expect to reduce the per capita water use to the limit of 100 L/capita-day which is apparently the lowest limit for a well functioning linear urban system (Falkermark and Widstrand, 1992) that can be reached by water conservation. Other cities use various degrees of water reclamation and reuse but start with a higher per capita water use reduced by reclamation of used water and stormwater.

A high density Qingdao ecoblock with 430–515 people/ha appears to be an anomaly which should be further researched as to the feasibility and sustainability of the concept regarding the used water reclamation. Qingdao's treatment of black water consists of 'sequential batch reactors' described in a promotional video (Green Dragon, 2008) as septic tanks, followed by wetland. Putting a surface flow wetland treating black water, even after pretreatment, in a densely populated ecoblock is not possible from health reasons in the US and other advanced countries. Based on the WEF (2001) a detached fenced off wetland would have to have a fully submerged flow. Based on the WEF (2001) manual the minimum area of the wetland serving 1500–1800 people will have to have an area of about $\frac{1}{2}$ hectare or one football field. Also such wetland will have a relatively large evapotranspiration during dry summer days period. Wetlands also emit methane and nitrous oxides.

Surface Drainage for Runoff and Clean Water. All ecocities use surface drainage for collecting urban runoff and clean water inputs (see Figure 13.3). All cities will use extensively best management practices for urban runoff such as pervious pavements for infiltration, capture and storage in underground basins, and its reuse for various purposes such as irrigation, fire protection, and some plan to tap into the groundwater resources for reclaimed water for nonpotable water supply. Use of green roofs has not been planned on a large scale with the exception of Hammarby Sjöstad.

Cost Per Unit. The cost per unit has been estimated by dividing the capital costs of the development by the number of dwelling units or by the population in the city divided by 2.5. The cost is only approximate and varies based on the standard of living of the country and a type of the buildings which determines population density. The lowest cost is for an apartment in the Qingdao ecoblock which consists of several high rises and buildings ranging from six to twenty four stories. The highest cost is in Masdar which has the highest degree of recycling, very high cost of desalination, and proposes a futuristic public transportation system. The affordability and standard of living also determines the cost. For example, the development of Sonoma Mountain Village was slowed down by the housing financing crisis in the US and that of Masdar is affected by the global financial crisis. A true convertibility between Chinese currency and US $ has not yet been established but the data for Tianjin are based on information from Singapore which is a major partner of the development. The Tianjin ecocity is located in the large Tianjin – Binhai new development area and the cost of the housing is commensurate with the middle class incomes of people living and working in this giant development which will contain an airplane (Airbus) manufacturing facility, research centers, universities, large port facility and many other industries located in 'green' environments.

Water Centric Development Opportunities. Hammarby Sjöstad, Dongtan, and Tianjin are clearly water centric whereby open water and canals are the architectural centerpieces of the development and will have an aesthetic role, provide recreation and local transportation. By locating their advanced wastewater treatment plant at the fringe of the city and directly discharging the used treated wastewater into the Hammarby Lake connected to the Stockholm Bay without water reclamation, the city has missed its opportunity for water reuse. Dongtan and Tianjin in China considered using the water bodies inside the city for discharge and additional treatment of reclaimed water. The desert city Masdar will apparently create small artificial streams transecting the city. Masdar also has sophisticated underground potable water, reclaimed water and used water conveyance systems (Hartman et al., 2010). Qingdao, Sonoma Mountain Village, and Treasure Island will not have permanent streams, natural or artificial, planned within the ecocity boundary. Sonoma Valley Village is planning to create habitat bioswales with wetlands for stormwater conveyance transecting the village and connected to a storage basin from which water will be reused. Qingdao created two conveyance systems for reuse: one for the reclaimed black water via a chain of wetlands, the other for stormwater both ending in an underground storage facility, followed by reuse. The architectural rendering of the Qingdao ecoblock does not show any surface stormwater conveyance to the central storage basin.

13.4.1 Lack of Macroscale Assessment

The analysis has revealed some problems with the lack of macroscale of some key components of the triple bottom line assessment. The literature data and personal inquiries revealed that the major companies providing engineering designs of the ecocities have or are developing models by which they calculate energy or water use by the city. However, a comprehensive environmental assessment model is not yet available. Such model or models are needed to test the following hypotheses:

- There is a quantifiable limit to water reuse at the macro, i.e. city scale. Sustainable urban water systems are characterized by a significant amount of recycling, supplemented by rainwater and/or make-up water import. This water import (and consequently export) required for flushing out conservative or poorly degradable contaminants within the system, constitutes a key macro-scale design parameter for the good health of people and aquatic biotic integrity. Quantification of the amount of import is critical for the design of these systems, including assessing the needs for ecologic flow to sustain viable aquatic life of the water bodies.
- Reuse may lead to accumulation and exceedance of thresholds for good water quality for several key water quality accumulative parameters in the surface streams, canals and lakes of the water centric ecocities such as nutrients (nitrogen and phosphorus) to prevent highly eutrophic and hypertrophic conditions. The hypertrophic status is characterized by dense blooms of toxins producing cyanobacteria and algae. This could happen in situations where the surrounding water bodies are already near such thresholds as might be the case of the Chinese ecocities of Tianjin and Dongtan.
- During the previous paradigm of building the underground urban water/wastewater conveyance, and treatment and disposal infrastructures based on the minimum capital and OMR costs, water reuse and energy recovery from used water and organic solids was not economical because it would have required another pipeline system and pumping to bring water and reclaimed energy back to the city. If the systems were broken into a number of semi-autonomous (with respect to water/stormwater/used water management) clusters (ecoblocks), water and energy reclamation and reuse would become socially attractive. Then there would be no need for large underground interceptors and pipelines; potable water would not be used for irrigation and providing ecologic flow to flow deficient urban streams or even recharging groundwater zones to prevent subsidence of historic buildings (e.g. in Boston). However, the optimal size of the cluster is not known and most likely there may be none as documented in Table 1 where the ecocity communities ranged from 5,000 to 500,000 inhabitants. Is the Qingdao ecoblock with 1800 people the most economical, efficient, and sustainable cluster? Or, is it the 50,000 people Masdar city/cluster or 500,000 Tianjin subdivided into clusters? The cluster size, population density within the cluster, ratio of green and built areas, quantity of water reused and renewable energy produced are examples of decision variables but there is no limit on the size as long as the hybrid distributed system is proposed (Novotny, Ahern and Brown, 2010). The future eco-communities may range from a cluster of houses, a large resort, a high-rise building with

hundreds to thousands of tenants, to large cities with partially or fully distributed water/stormwater/used water system.

13.5 SUMMARY

Currently, there are dozens of urban developments throughout the world claiming to become 'an ecocity'. A few, with various degrees of success, are striving to become certified as 'One Planet Living' community. In our analysis we have compared only a fraction of the most publicized developments. Because of their frugality with respect of energy and water requirements, the 'new cities' can be built in 'hostile' environments such as arid areas with poor soils (Tianjin, Masdar) or decontaminated brownfields (Hammarby Sjöstad in Sweden, Dockside Greens in British Columbia, 2021 Olympic sites in London, Treasure Island in San Francisco). This may relieve the pressure on valuable agricultural lands, wetlands and forests even in countries still undergoing excessive population growth. Building new cities in fragile natural environments such as a desert should still take into account the natural systems that need to be protected.

Masdar is an unusual case that will be a test ground for the Cities of the Future research and development. It is showing that the state of the art of building cities has reached a point where there are almost no limits but at a cost which still may be prohibitive elsewhere. The main source of water in Masdar is evaporation/condensation desalination, powered by abundant solar energy, of Gulf of Persia (Arabia) water which has high salinity.

In developed countries, the direction will be more towards retrofitting the existing cities and reducing or even reversing urban sprawl by bringing people back from distant energy gobbling low density suburbs to the retrofitted and water and energy efficient cites. Unfortunately, in old municipalities, bringing new sustainable concepts into rebuilding and retrofitting may be running into resistance and obstacles caused by existing regulations and traditions.

ACKNOWLEDGEMENT

Partial support for the work described in the article was provided by contracts from US EPA and WERF (Steve Moddemeyer) and from CDM, Inc. which is greatly appreciated. The views expressed in the article are solely those of the authors.

REFERENCES

Daigger, G. (2009) Evolving urban water and residuals management paradigms: Water reclamation and reuse, decentralization, resource recovery, *Water Environment Research* **81**(8):809–823.

Falkermark, M. and C. Widstrand (1992) "Population and water resources – A delicate balance," *Population Bulletin* **47**(3):1–36.

Green Dragon (2008) www.greendragonfilm.com/qingdao_ecoblock_project.html.

Hartman, M., M.B. Knell, and J.W. Witherspoon (2010) Masdar City's integrated approach to sustainability, Proc. Cities of the Future 2010 Conference, March 7–10, Cambridge, MA. pp. 104–117.

Hill, K. (2007) Urban ecological design and urban ecology: an assessment of the state of current knowledge and a suggested research agenda, in *Cities of the Future: Towards integrated sustainable water and landscape management* (V. Novotny and P. Brown, eds), IWA Publishing, London, UK.

Hoekstra, A. Y. and A. K. Chapagain (2008), Globalization of Water: Sharing the Planet's Freshwater Resources. Oxford: Blackwell Publishing.

Kennedy, C., J. Cuddihy, and J. Engel – Yan (2007) The changing metabolism of cities, *Journal of Industrial Ecology*, **11**(2):43–59.

Novotny, V. (2008) Sustainable urban water management, In *Water & Urban Development Paradigms (J. Feyen, K. Shannon, and M. Neville, eds.)* pp. 19–31, CRC Press, Bocca Raton, FL.

Novotny, V. (2010) Water and energy footprints of the Cities of the Future, Invited key note presentation at the 2010 Singapore International Water Week June 28-July 2, published in *Water Policy and Technology* **4**(5).

Novotny, V., and E.V. Novotny (2009) Ecocity – Evaluation and Synthesis, www.aquanovallc.com.

Novotny, V., J.F. Ahern, and P.R. Brown (2010) Water-Centric Sustainable Communities: Designing, Building and Retrofitting the Next Urban Environments, J. Wiley & Sons, Hoboken, NJ.

O'Riordan, J., W.P. Luccey, C.L. Barraclough, and C.C. Corps (2008) Resources from waste: An integrated approach to managing municipal wate and waste systems, *Industrial Biotechnology* **4**(33):238–245.

Prince George's County (1999) *Low-Impact Development Design Strategies: An Integrated Design Approach*, http://www.epa.gov/owow/nps/lid/lidnatl.pdf (accessed September 2010).

Register, R. (1987) *Ecocity Berkeley: building cities for a healthy future*, North Atlantic Books. ISBN 1556430094.

USGBC (2005) *Green Building Rating System for new Construction & Major Renovations Version 2.2*, US Green Building Council, Washington, DC., http://www.usgbc.org.

USGBC (2007) *LEED for Neighborhood Development Rating System, Pilot Version*, US. Green Building Council, Washington, DC., http://www.usgbc.org.

Water Environment Federation (2001) *Natural Systems for Wastewater Treatment*. Manual of Practice FD-16, Alexandria, VA.

WWF (2008) One planet living, http://www.oneplanetliving.org/index.html.

Part IV

Leading by Example

Chapter 14

Water, neighborhoods and urban design: micro-utilities and the fifth infrastructure

Vicki Elmer[1] *and Harrison Fraker*[2]

[1]*School of Architecture and Allied Arts, University of Oregon*

[2]*College of Environmental Design, University of California, Berkeley, California, USA*

ABSTRACT

Global warming has given a new dimension to urban design efforts that seek to integrate the infrastructure systems of a city into a sustainable and more natural built environment. The result has been a transformation in approaches taken by leading edge design practitioners and the elites responsible for urban infrastructure systems. Many group these responses under the rubric of the 'eco-city.' Nowhere has this been more evident than in the water and energy sectors. Mitigating carbon emissions has been taken up by local governments around the world, while water is the most visible vector for the effects of global warming.

This paper first describes innovations in energy, solid waste and wastewater at the building level before examining how they are combined in six European neighbourhoods as an integrated 'micro utility,' taking particular advantage of the public and private landscape. We conclude that greater integration of waste systems and the urban landscape is the final secret to achieving 'zero emissions' operation. The neighbourhood is the scale that makes more sustainable utility systems possible and the landscape is both the technical underpinning and the aesthetic place making element, becoming the fifth infrastructure.

Keywords: Integrated water management; landscape; eco-city; eco-block; urban design; water reuse; decentralized wastewater treatment; sustainable development; global warming; urban planning

14.1 INTRODUCTION

Global warming has given a new dimension to urban design efforts that seek to integrate the infrastructure systems of a city into a sustainable and more natural built environment. For several decades, architects, planners, and designers have been calling for a more compact urban form that integrates nature with the city, as well as greater use of energy efficient building and transportation alternatives. However, the need to mitigate and adapt to climate change has created a new and more urgent driver for change. The result has been a transformation in approaches taken by leading edge designers and the elites responsible for urban infrastructure systems. Nowhere has this been more evident than in the water and energy sectors (Daigger, 2009; Dreiseitl et al., 2009; Hermanowicz, 2008 Lindsay, 2010; Poetz and Bleuze, 2008; Woltjer and Al, 2007). Mitigating carbon emissions has been taken up by local governments around the world, while water is the most visible vector for the effects of global warming. Many describe these responses under the rubric of the 'eco-city' (see Figure 14.1).

Figure 14.1 EVA Lanxmeer eco-city with gray water filtered in wetlands, solar panels and energy efficient buildings. Photograph Courtesy of Herbert Dreiseitl and Atlier Dreiseitl

The eco-city rejects the idea of waste, instead seeking to transform it into beneficial uses within the city. In so doing, it seeks to reduce inputs of water and energy from afar. This concept is behind efforts to decentralize the production of energy and food and to create more local 'green' employment. It also powers the three 'R's' of solid waste management as well as transportation efforts such as transit oriented development, walkable neighbourhoods, and 'complete streets.' Architects, planners and landscape architects have used many of these innovations at the building site level over the past twenty years. The next breakthrough is to integrate these systems at the neighbourhood, block or development cluster to begin to approach zero emissions.

This paper begins by describing energy and water innovations at the site and building level before examining the experiences of six cities that have tried to integrate water, energy and solid waste utilities in new neighbourhoods. In so doing, these pioneers have begun to move towards the development of micro-utilities that use the landscape as the fifth infrastructure for true sustainability.

14.2 BUILDING LEVEL SOLUTIONS

14.2.1 Energy: conservation & on-site production

Conservation. Over the past 40 years, since the first Earth Day in 1970 and the subsequent oil embargo in 1973, important work has focused on energy conservation in buildings. The R&D conducted on 'super insulation', air infiltration, passive solar energy, shading, natural ventilation, daylighting, and energy efficient lighting and appliances forms a solid foundation on which we now confront the challenges of climate change. Study after study has shown when these strategies are designed and deployed wisely, they can save from 40–60% of energy consumption in typically constructed buildings (McKenzie & Company, 2009).

Smart and Decentralized Energy Production. Recent developments in technology have achieved even greater efficiencies in the area of lighting with LED's and in 'smart buildings' with wireless lighting controls, occupancy sensors and real time HVAC management systems. Materials research and manufacturing breakthroughs in photovoltaic (PV) production are on the verge of making the wide spread application of PV's on buildings cost effective. Even now, PV's are cost effective in some markets, such as California. With this trajectory in technical development, the idea of designing 'zero carbon' buildings is a tantalizing goal for the not so distant future.

14.2.2 Water

During the past twenty years, designers and planners have been devising solutions to decrease the demand for water at the building and site level and to prevent flooding. The following innovations can reduce water demand in residential buildings in the developed world by 50 to 85%.

Gray Waste Water[1]. Reuse of gray water at the building and site level has also emerged over the past 20 years as an important tool in reducing potable water demand. The Pacific Institute estimates that half the water used in homes in the United States can be reused for irrigation and toilet flushing. On-site gray

[1]Because regulations differ from place to place, so do gray water definitions. Generally gray water, or light gray as some call it, refers to water from wash basins, laundry, and showers; dark gray includes kitchen wastewater, although some include kitchen waste water in the general designation of 'gray.'

water systems for toilet flushing are common outside the United States (Allen, et al., 2010).

Black and Brown Waste Water[2]. This is an area where small scale technologies are advancing rapidly. This concept processes the wastewater on-site and reuses the water for toilet flushing and irrigation. One approach to treatment uses natural processes such as the Eco-Machine, the Living Machine or specially engineered 'boutique' systems. These systems flush the effluent through a series of tanks where living plants clean the waste and also take up the metals (Lens et al., 2005; Steinfeld & DelPorto, 2007). A second approach uses Membrane BioReactors (MBRs) to filter toilet waste through very small holes in a membrane. MBRs can filter out viruses, pharmaceuticals and even metals. However, as the size of the particle to be filtered out diminishes, the amount of energy needed to process the wastewater increases. The miniaturization of MBR's, rapidly declining costs and increasing energy efficiency have resulted in robust small scale package wastewater treatment plants that can operate with a minimum of maintenance (Asano et al., 2008; Judd, 2006).

Stormwater. Rainwater harvesting is an important water source at the building level, and in many cases could provide anywhere from half to over 200% of the water needs of a building, or a city. In addition, as global warming results in heavier floods and more severe droughts, the natural cycle of water percolating into the soil or evaporating into the atmosphere in urban areas is disrupted by roofs, impervious pavements, and by channelling storm water into underground sewers (see Figure 14.2). Buildings and their sites are increasingly being designed to slow down the rush of storm water. Such efforts include making driveways and parking lots more pervious, the use of swales instead of storm sewers, rain gardens and tree planting. Green roofs and walls are tools which also reduce the urban heat island effect and energy use (Novotny et al., 2010).

14.2.3 Integrated systems for water, waste & energy

Pilots for Integrated On-Site Concept. The most sustainable water, wastewater, and solid waste concepts have been developed in Europe and in the developing world. Pilot projects based on this concept attempt to mimic a natural closed loop cycle. Black, yellow and gray waters are separated at the building level, treated and reused with nutrient and energy recovery. Many pilots began as an effort to recover nutrients from black and yellow water, but as their sights turned to the production of energy through biogas and methane production, it became quickly evident that human biosolids combined with food waste was a more effective alternative (Elmer, 2009).

[2]There is no uniform convention for designating brown and black water. Brown and black water can be the effluent from a source separation toilet containing only the feces. Black water also refers to water from toilets with both urine and feces.

Figure 14.2 Comparison of storm water runoff in rural, agraland, and urban areas. Photograph Courtesy of Herbert Dreiseitl and Atlier Dreiseitl

The GTZ (Deutsche Gesellschaft für Technische Zusammenarbeit—now part of GIZ) headquarters building in Germany separates gray water, which is processed in wetlands. It also has source separation toilets which permit faeces to be mixed with food scraps for biogas production while urine is used as a fertilizer. Other buildings in Europe such as the BedZed development in London's Dockside area have partially integrated systems.

14.3 NEED FOR NEIGHBOURHOOD SCALE SOLUTIONS

The individual flows of energy, water and waste at the building and site level are small. They are constrained by the particulars of their site and climate, and limited by their program of operation which determines the size and timing of their energy and water use. In addition, when considered as a stand alone design problem, important external energy and water flows on which buildings depend are not included. Important elements for the liveability of a building such as adjacent water ways cannot be dealt with at the site level. Finally, solid waste, usually dealt with at a city or regional level, turns out to be an essential element in processing wastewater into energy and nutrients. From this perspective, while buildings are an important component in achieving a 'zero emissions' eco-city, there are larger system concerns which suggest that opportunities for creating sustainability need to be examined at a larger scale.

Neighbourhoods and intermediate to large scale mixed-use development projects offer intriguing advantages and opportunities. First, neighbourhoods have a mix of uses which makes it easier to balance loads and match the intermittent supply of renewables. Second, neighbourhoods have larger flows of energy, water and waste with which to work. Third, their design can influence transportation choices and reduce automobile use. Fourth, when designing at the neighbourhood scale, the urban landscape can be brought into play to temper the climate, absorb carbon, clean stormwater and sewer effluent, provide biomass for energy and even grow food (see Figure 14.3).

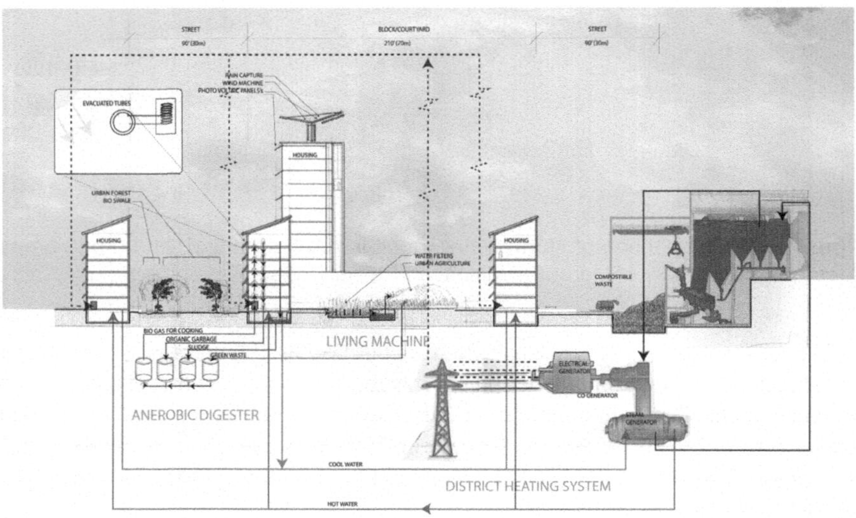

Figure 14.3 Schematic of Sustainable Infrastructure Systems for an Eco-Neighbourhood or Eco-Block. Source: Harrison Fraker and team. 2010, University of California

Neighbourhood scale development (from 400 to 10,000 units of housing) is also a relatively typical form of development, both for private real estate developers, but also for cities doing urban renewal on underused or 'brown field' industrial sites all over the world. If neighbourhoods can become their own micro-utility, supplying most if not all of their energy while treating and recycling their water and waste, a whole new form of sustainable development is possible. In addition, if all of the energy can be generated locally and much of the waste processed on site, the cost and loss of efficiency in distribution infrastructure and transport can be avoided.

14.4 SIX ECO-NEIGHBOURHOODS

We identified six eco-neighbourhoods built in European cities which exemplify the whole systems approach to urban design: **Bo01/Western Harbor** in Malmo and

Hammarby in Stockholm, Sweden; **Kronsberg** in Hannover and **Vauban** in Freiberg, Germany; **Lanxmeer** in Culembourg, NL; and **solarCity Pinchling** in Linz, Austria (see Figure 14.4). There are other developments which look at water/energy/waste holistically which have been designed but not built (Qingdao, China; Denver's Living City Block; Portland's proposed Eco-Districts and others noted in the Novotny and Novotny article in this volume).

Figure 14.4 Aerial photograph showing the urban plan of **solarCity Pinchling** in Linz Austria. Source: Linz AG website.

These neighbourhoods fulfil basic criteria for sustainable design: 1) each is transit-oriented with an emphasis on walking and biking and convenient access to public transit (within 5 minutes). 50–80% of all daily trips are by pedestrian, bike or public transit, 2) all have aggressive conservation goals of approximately 100 kWh/m2-year for total energy use and employ local renewable energy strategies, 3) all have attempted closed loop systems for water/wastewater and have integrated natural water features into the design of the community; 4) all are mixed use with a jobs to housing balance of at least 50%, 5) all have a net density of approximately 30–40 units per acre and range in size from 400 to 8000 housing units; and 6) most have been developed with a high degree of citizen participation; 7) many of these feature local gardens for on-site food; and 8) solid waste is integrated in some for energy production.

14.4.1 Energy in the eco-neighbourhoods

These neighbourhoods all use passive solar energy for the building location, design, insulation, siting and construction. They also develop sustainable energy resources on and off site before using conventional energy sources. Together, they demonstrate four possible strategies for generating energy from local renewable sources: wind, solar, geothermal and waste, each with a different emphasis and combination. **Bo01/Western Harbor** uses local wind generation to power a geothermal ground and ocean water heat pump for heating and cooling. **Hammarby** has 3 different waste-to-energy systems: the first burns combustible garbage to power a local district heating and electric cogeneration plant (off-site), the second recovers heat from the sewage treatment system and the third converts sludge to biogas for cooking (1000 units) and to power local buses. Similarly **Lanxmeer** has designed an on-site biogas system that uses its blackwater, garden and kitchen waste, which is owned by the community (see Figure 14.5). **Kronsberg** has two large scale wind machines (totalling 3.2 megawatts) which generate 50% of its electricity combined with a gas fired heating and electric cogeneration plant, which provides the other 50% electricity. **Vauban** has a local heating and electric cogeneration plant powered by waste wood chips from the area.

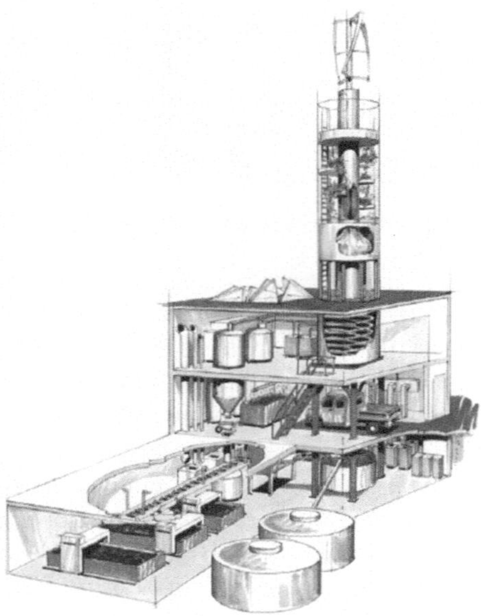

Figure 14.5 EVA Lanxmeer Biogas Schematic. Source: www.lanxmeer.nl

While all these neighbourhoods demonstrate good energy conservation standards, Kronsberg, Vauban and solarCity have many sections that meet the very aggressive 'passive house' standard of 15 Kwh/m2-y for heating. Vauban also has several solar developments (called 'plus energy houses') which combine a passive solar direct gain system for heating and a roof-top photo voltaic array for electricity, that delivers +15% energy back to the city. Together, the neighbourhoods have employed all types of solar collection. Bo01 uses evacuated tube collectors to assist the district heating system. Hammarby uses flat plate panels and evacuated tubes to preheat domestic hot water. As a test case, Kronsberg combines a large solar hot water array with a large seasonal storage tank in order capture summer solar energy to augment winter solar heating. All the neighbourhoods have applied photo voltaic arrays.

14.4.2 Water and waste in the eco-neighbourhood

All the eco-neighbourhoods collect storm water from the streets to filter and clean it using different forms of bio-swales, before returning the storm water to streams, lakes and/or the aquifer. One innovative approach is the use of 'green' transit tracks. There are some green roofs in Hammarby, Bo01, Lanxmeer, and Kronsberg which help to buffer rainwater and act as a form of insulation both winter and summer. Mandatory tree planting is required in some for storm water. The goal for most of the communities is to manage the water so that the areas' natural water balance is the same after development as it was before.

All of the developments maximize the amenity value of water. They use water features to form or preserve the landscape, and as recreational and ecological preservation areas. The best incorporate features that are aesthetically and spiritually pleasing. Some of the developments have used engineered features while others have restored or incorporated ponds, creeks and waterways otherwise into the urban design (see Figure 14.6).

Many of the developments have used the landscape to process gray water sewage and other ecological sewage systems. A building in Vauban transports faeces through vacuum pipes to an on-site facility which combines this with foodscraps to generate biogas for cooking. Lanxmeer does not send any of its sewage to the city's wastewater treatment plant. Instead, buildings in this development have separate pipes for gray and black (toilet) wastewater, with the black going to the biogas plant noted above, and the gray water purified in reedbeds and ponds (Poetz, and Bleuze, 2008). solarCity Pinchling–Linz has piloted an urban urine diversion toilet system in a primary school and for 88 dwelling units. The urine is currently disposed into the central system due to the lack of permits for its use in surrounding agriculture (Ulrich, 2009). This development also uses wetlands to process graywater.

Stream and Creek Restoration. Another component of the eco-city is stream and creek restoration. This is important to assist building level storm water solutions and to promote infiltration, increase wildlife diversity and the amenity value of the natural water way. The eco-neighborhoods above have restored water bodies and streams. In addition, Zurich, Copenhagen, the eco-city in Tianjin (see

Figure 14.7) are in the process of restoring their canals to natural streams. In South Korea, a large freeway constructed in the bed of a former river was dismantled and the river restored – to the delight of residents when the dragonflies returned to the area.

Figure 14.6 Walkway and storm water in Bo1, Malmo, Sweden. Photograph by Brandon Schauer http://brandonschauer.com

Figure 14.7 Creek Restoration in Tianjin, China. Photograph Courtesy of Herbert Dreiseitl and Atlier Dreiseitl

Solid Waste and Recycling. All the eco-cities or clusters have well developed systems for solid waste collection and recycling. Bo01 and Hammarby use an evacuated (vacuum) tube system. In Bo01 above ground receptacles for recyclabes lead to underground pipes which transport the material outside the area. Trucks then pick up the material and take it for energy production. In addition, food scraps for 200 households are processed with a garbage disposal (food grinder) to produce methane that is used to power school buses. In both the Swedish developments dried organic waste is burned in a co-generation plant, while green waste is composted. The other case studies all have aggressive recycling goals of zero waste and compost green waste.

14.5 IMPLICATIONS FOR URBAN DESIGN AND WATER

These neighbourhoods show that zero emissions operation is achievable on a decentralized basis. They also show that it is not just a matter of finding and applying the 'right' technical systems and following the 'right' development process, as important as these may be. It involves thinking of technical strategies and urban design as one, creating a high quality built environment, one which fosters a vibrant experience. The challenge for designers is to learn how sustainability strategies can enhance the quality of the built environment and deepen the experience of peoples' everyday lives. How do concepts of urban design – the design of the streets, blocks, parks and urban landscape interact with strategies for sustainability? Are there conflicts? What, if any, trade-offs have been made?

On one level, the urban design, the principles of urban form for these neighbourhoods are similar (see Figure 14.8). They assume a traditional plan of streets and blocks. Each plan is then modified to take advantage of the particular conditions of the site and landscape, including such features as lakes, shorelines, hills, orientation for sun and wind, and views. Different open space strategies for parks, recreation areas, courtyards, plazas and urban landscape functions further enrich the form of each neighbourhood. While quite traditional as an urban design framework, the subtle responses in design of the blocks, the architecture, the streets and especially the design of the urban landscape are where the neighbourhoods come alive, suggesting further design strategies for sustainability.

An additional feature of the integrated design approach is that the developer can become a micro–utility and could charge residents for the service, recovering the added cost of the systems. The micro-utility can then be turned over to the residents or the local jurisdictions for operation. An analysis of the cost of the integrated system was performed for the proposed eco-block in Chengdu by one of the authors. Sustainability initiatives were estimated to increase cost of development by 5%–10%. The 10 year internal rates of return (IRR) range from 2% to 10% depending upon the scenario. Payback periods range from 10 years to 7 years. The information control systems needed to coordinate the energy, water and waste flows are cost effective and not prohibitively expensive (Fraker et al., 2008).

Figure 14.8 Schematics of urban design patterns for Kronsberg, and Vauban, Germany. Source: Harrison Fraker and team. University of California. Shaded area is ¼ mile diameter

These efforts also point to a new approach to the design of public open space, including the street right-of-way and local parks. In many jurisdictions, 30–40% of urban land is public open space. In the private sector, even in high density neighbourhoods (100+ units per acre, net) most site designs provide more than enough area in the semi-private courtyard space for wastewater treatment. If the area inside private property lines, not covered by a building foot print, is included, more than 50% of urban land is available to support ecological functions. This presents a tremendous design opportunity for recycling water, urban agriculture and creating urban forests in both the public and private realms.

The next generation of water, wastewater and solid waste thinking integrates the urban landscape into the whole-system operation. The landscape can clean and retain storm water and clean sewage effluent as well using a 'living machine' or wetlands concept after the sludge has been removed and sent to the anaerobic digester. Or it can accommodate a small scale MBR. The area necessary to treat the effluent is easily accomplished within each block design. This enables the treated water to be collected and recycled at the block scale, eliminating the need for sewer lines and reducing the length of water supply lines. An extensive tree planting design strategy yields wood chips, which, along with other landscape clippings, adds fuel to the combustible solid waste-to-energy system. The urban landscape can also provide local food for residents along with tempering the climate, while enhancing the experience of the public realm.

The next generation of water-energy-waste infrastructure can also be used to mitigate the heat from climate change and the additional flow of water during wet periods. Figure 14.9 shows an integrated landscape that has the ability to create a micro climate of its own. In an orchard with processed gray water, the temperature is 80 degrees compared to over 100 at the entrance by the asphalt roadway. See also Figure 14.10 of Potzdammer Platz in Berlin, where local climate effects were consciously sought by the landscape architect.

Figure 14.9 Photograph by Tom Conelly, www.pbase.com. 2009

Figure 14.10 Potzdammer Platz, Berlin, Germany. Photograph Courtesy of Herbert Dreiseitl, Atlier Dreiseitl

14.6 CONCLUSION

The neighbourhoods described above indicate that a greater integration of waste systems and the urban landscape are the last pieces in the puzzle, the final secret to achieving 'zero emissions' operation. The neighbourhood is the scale that brings them into play and the landscape is both the technical underpinning and the aesthetic place making element, becoming the fifth infrastructure. The challenge will be to engage the centralized utilities and the technical wizards with the urban design professionals to make 'sustainability' a compelling whole. This will require new institutions and new ways of interacting. The goal must be to turn the city and its water into what Professor Elizabeth Meyer, Professor of Landscape Architecture at the University of Virginia has called 'sustaining beauty.'

We also recall that:

> *'Water is the heart of all life. In the past, we built water and wastewater systems infrastructure to protect ourselves from diseases, floods, and droughts. Now we see that fundamental life systems are in danger of collapsing from the disruptions and stresses caused by this infrastructure. New and evolving water technologies and institutions that mimic and work with nature will restore our human and natural ecology across lots, neighborhoods, cities and watersheds.' (Nelson, et al., 2007).*

REFERENCES

Allen, L., J. Christian-Smith, and M. Palaniappan, (2010). *Overview of Gray Water Reuse: The Potential of Graywater Systems to Aid Sustainable Water Management.* Oakland California: The Pacific Institute. www.pacinst.org.

Asano, T., F. Burton, H. Leverentz, R. Tsuchihashi and Tchobanoglous, G. (2008). *Water Reuse: Issues, Technologies, and Applications*, Metcalf & Eddy, New York.

Cervero, R. (1998). *The Transit Metropolis: A Global Inquiry.* Island Press: Washington, D.C.

Daigger, G. (2009). *Evolving Urban Water and Residuals Management Paradigms: Water Reclamation and Reuse, Decentralization and Resource Recovery.* Water Environment Research, Volume 81.

Dreiseitl, H. and D. Grau, K.H.C. Ludwig, eds. (2009). *Waterscapes: Planning, Building and Designing with Water 3rd Edition.* Basel, Switzerland: Birkhaeuser Verlag AG.

Elmer, V. (2010). *Blue-Green Waste Water Technologies*, the Sage Series on the Green Society, *Volume 10 Green Technology*, Editor(s) Robbins, Paul, and Dustin Mulvaney, J. Geoffrey Golson.

Fraker, H., students and ARUP, 2008. University of California at Berkeley. Powerpoint presentation at UC Berkeley's Institute of the Environment, http://bie.berkeley.edu/

Hermanowicz, S. (2008). *Sustainability in water resources management: changes in meaning and perception.* Sustainability Science, Springer.

Langenbach, H., J. Eckart, and G. Schroeder, (2008). *Water Sensitive Urban Design—results and principles.* 3rd SWITCH Scientific Meeting. switch@hcu-hamburg.de

Lens, P., G. Zeeman, and G. Lettinga, Eds., (2005). *Decentralised Sanitation and Reuse: Concepts, Systems and Implementation*, Water Intelligence Online © IWA Publishing.

Lindsday, G. (2010). *Peter Calthorpe's Vision for California: Save the Cities, Save the World* in Fast Company May 21, 2010. www.fastcompany.com

Krause, A. and A. Sayani, (2005). *Planning Sustainable Communities: Case Studies, Kronsberg.* in Planning New Communities, Places and People. On-line.

Judd, S. (2006). *The MBR Book, Principles and Applications of Membrane Bioreactors for Water and Wastewater Treatment.* Elsevier, Oxford.

Nelson, V., S. Moddemeyer, and P. Schwartz, (2007). *The Baltimore Charter for Sustainable Water Systems.* http://sustainablewaterforum.org/baltimore.html

Novotny, V., J.F. Ahern, and P.R. Brown, (2010). *Water-centric Sustainable Communities: Planning, Building and Retrofitting Future Urban Environments*, J.Wiley & Sons, Hoboken, New Jersey.

Malmqvist, P., G. Heinicke, E. Korrman, T. Bergstrom, and G. Svensson, (2006). *Strategic Planning of Sustainable Urban Water Management*, IWA Publishing, London, (p. 161-173).

McKinsey and Company (2009). *Unlocking Energy Efficiency in the US Economy.* www.mckinsey.com.

Poetz, H. and P. Bleuze, (2008). *Vorm geven aan stedelijk water.* Sun Publishing, NL. www.water-in-zicht.nl.

Steinfeld, C. and D. Del Porto, (2007). *Reusing the Resource: Adventures in Ecological Wastewater Recycling*, Concord, Mass: Ecowaters Books, www.ecowaters.org.

Woltjer, J. and N. Al, (2007). *Integrating Water Management and Spatial Planning.* Journal of the American Planning Association, Vol. 73, No. 2. (p. 211–222).

Ulrich, L. (2009). *Urban Urine diversion & greywater treatment system, Linz, Austria.* GTZ ecosan program – ecosan@gtz.de, www.susana.org.

Chapter 15

Keys to successful transitioning-lessons from the Netherlands' and Japanese delta cities

Rutger de Graaf[1,2] and Frans van de Ven[3,4]

[1]*DeltaSync, Delft, the Netherlands*
[2]*Rotterdam University of Applied Sciences, Rotterdam, The Netherlands*
[3]*Delft University of Technology, Faculty of Civil Engineering and Geosciences, Urban Water Management, Delft, The Netherlands*
[4]*Deltares, Urban Water Management, Delft/Utrecht, The Netherlands*

ABSTRACT

Urbanization, climate change and social developments increase the vulnerability of cities.

This chapter provides an outline and the features of future water cities. A number of properties of future watercities are discussed and illustrated with examples from Japan and the Netherlands. Future water cities are sustainable, climate robust, adaptable, healthy and pleasant. Reliable technologies are available to achieve this. Mainstreaming them would be the next logical step. Two key conditions for mainstreaming of these urban water management innovations are presented in this chapter. The first condition is including urban water management innovations in spatial planning and development. The second condition is stakeholder receptivity to urban water management innovations. These key conditions are discussed and illustrated with examples from the Netherlands and Japan.

Keywords: Transitions; urban water innovations; Japan; the Netherlands; receptivity; urban planning; resilience; climate adaptation; vulnerability

15.1 INTRODUCTION

Rapid urbanization is the most significant geographical development of the past century, both in the Netherlands and Japan. On a global scale, the proportion of people living in cities now exceeds the rural population (UNFPA, 2007). This urbanization predominantly takes place in delta areas, coastal areas and flood plains, areas that are exposed to flood risks. In 2030, some 50% of the world's population is expected to be living within 100 kilometres of the coast (Adger et al., 2005). The advantages of living in a delta region, a coastal or a river flood plain seem to outweigh the geophysical disadvantages such as flood risks and soft soils.

The growth of cities in combination with the industrialization of our economy brought us to the edge of feasible and sustainable urban developments. Negative effects of urbanization on urban water systems include (Team Urban Water Management, 2009):

- Parasitic behavior;
- Vulnerability to extreme weather conditions:
  - Flooding
  - Drought and water shortages
  - Heat
- Land subsidence
- Land scarcity

These effects are aggravated by the continuous economical, technological, physical and social developments in urban areas. Due to these developments the city requires continuous adaptation. These adaptations which cannot be offered by the existing, rigid urban system. Decay is often the result, until such an urban area is upgraded again in a radical urban renewal project.

Nowadays, the ambition of urban development and redevelopment is to create an adaptable, sustainable, climate-resilient, healthy, and pleasant environment for living and working, while supporting a diverse ecological system. This holds for urban development all around the world, including the Netherlands and Japan. The crucial question is: How to achieve this? The answer to this question is neither a merely technological innovation, nor governance-, process- or organisation-oriented innovations. Many interacting innovations in multiple domains are required to obtain a successful transitioning process. In this chapter, we will describe some examples from the Dutch and Japanese urban water management practice.

15.2 WATER, THE KEY TO IMPROVEMENT

Tackling the aforementioned problems and taking up the ambitions in delta areas and flood plains requires that water is put at the very heart of urban development. Water underlies the delta cities; that is why we call them Water Cities (Hooimeijer et al., 2005). In water cities, planning and design of the water system – in relation to soil and subsoil – starts even *before* the urban planning and design process. Urban design and the civil engineer's design of the drainage system and water management system are integrated in a co-creative process. The introduction of the Water Assessment Procedure (Landelijke Projectgroep Watertoets, 2003; Van Dijk, 2008) and the GeoCheck procedure (Van Beek and Hounjet, 2009) illustrate this integration in the Netherlands.

The local water system provides the backbone and the grid for the urban design, allowing urban functions and functionalities to be located on specific sites and setting boundary conditions for architectural and landscape design. Organizing the urban water system - surface water, groundwater, storm water, drinking water, and wastewater - is therefore the first and essential step in creating a high quality urban environment. Examples of such new urban areas in the Netherlands are Leidsche Rijn in Utrecht (Figure 15.1), Stad van de Zon (City of the Sun) in Heerhugowaard, EVA-Lanxmeer in Culemborg, Nesselande in Rotterdam (Figure 15.2), IJburg in Amsterdam, the Noorderplassen in Almere and De Groote Wielen in 's Hertogenbosch (Figure 15.3).

Now let us take a closer look at how today's ambitions of urban development – that is to create a sustainable, climate-resilient, adaptable, healthy, and pleasant urban environment for living and working, while supporting a diverse ecological system – are being implemented in the Netherlands' and Japanese water cities.

Figure 15.1 Floating houses in Leidsche Rijn, Utrecht, the Netherlands: a flexible and adaptable way of urbanization (Photograph: W. Lindemans)

15.2.1 Sustainable

Sustainability is a rather ambiguous demand. Rijsberman and Van de Ven (2000) demonstrated that there are four fundamentally different understandings of sustainability. This resulted in four approaches to developing sustainable plans for urban water management. A recent elaboration of what they called the Socio Approach to sustainability was made for the water management of the city of Almere. This city formulated seven principles based on the cradle-to-cradle philosophy (Braungart and McDonough, 2002) 'for an economically, socially and ecologically sustainable future, as an inspiring guide for everyone involved in the decades to come in the design of Almere as a sustainable city' (Gemeente Almere, 2008).

Figure 15.2 Living with water in Nesselande, Rotterdam (source: Municipality of Rotterdam)

Another approach to sustainability is *water sensitive urban design*, developed by Australian scientists as one answer to the need for sustainable urban design in Australia (Wong and Eadie, 2000; Mouritz, 2000; Lloyd et al., 2002). They provided new solutions for stormwater management and diverse water supply sources that are integrated in the urban design and strengthen urban landscape quality. Rijsberman et al. (2000) would call this an example of an Eco-approach.

From a more Ratio- or Carrying Capacity – approach to sustainability, we first of all have to make urban areas less parasitic, i.e. less dependent on external resources and less waste-producing. In order to achieve this we should reduce the environmental footprint of cities (Team Urban Water Management, 2009):

- Reducing the inputs of water, food and energy by recycling water, nutrients and heat

- Reducing imports of sand, earth and building material
- Multifunctional use of water
  - Using surface water as space for urban development
  - Using water as an energy source for heating and cooling
  - Using the available water for food production
  - Using water to prevent land subsidence

Figure 15.3 Paleiskwartier in 's Hertogenbosch, the Netherlands, the surface water on top of a parking garage functions as a heat collector for the apartment building (Photograph: E. Aparicio)

We have to make better use of the internal water, food and energy resources and we have to increase efficiency. The available water – surface water, groundwater, storm water, drinking water and wastewater – is to be utilized more actively as a resource, rather than only be managed and controlled. The final objective is to reduce the environmental footprint of the urban area by creating a more closed city.

Many pilot projects in the Netherlands and in Japan demonstrate that this more closed city is no longer an utopia, examples include:

- In the Paleiskwartier in 's Hertogenbosch a pond is used to collect solar heat in summer. Heat from this pond is stored in an aquifer thermal energy storage and is used in winter to heat the houses, offices and apartments in the area
- A significant part of the effluent from the Odaiba wastewater treatment plant in Tokyo is polished and used for toilet flushing and irrigation in the urban area in the vicinity

- Storage capacity in surface water requires a lot of space. This space can be used for floating housing or housing on piles
- Local mining of sand for filling of buildings sites and roads. This mining creates a lake for stormwater retention, for recreation, floating housing and so on. An example is the Centrale Plas in Almere (Figure 15.4)
- Food production on green roofs in Japan (Figure 15.5)

Figure 15.4 Centrale plas Almere (Photograph: Alex Verweij)

15.2.2 Climate-resilient

Climate change is expected to cause more extreme weather conditions more often. That means more frequent and heavier rainfall, longer lasting droughts and more frequent extremely high temperatures. It seems wise to start protecting ourselves against these conditions by taking adaptive measures, in particular no-regret, cost-effective measures (Van Nieuwkerk et al., 2010). Van de Ven et al. (2009) showed that at least 180 technical and non-technical measures can be included in urban design and management to strengthen the water-resilience of an urban area. This makes climate resilient urban development a selection and design problem, rather than a technological innovation challenge. An interesting strategic guiding model for making such a selection from this long list of optional measures is provided by theory on how to reduce vulnerability:

Figure 15.5 Food production on green roofs in Japan

Four system capacities need to be strengthened in order to reduce the vulnerability of the urban environment (De Graaf et al., 2007). These are:

Threshold capacity,	measures to prevent damage, e.g. levees, pumps;
Coping capacity,	measures to reduce damage during flooding and droughts, e.g. mounds, local reservoirs, evacuation drills;
Recovery capacity,	measures to reduce indirect losses by repairing the damage, e.g. legislation to make labour capacity available;
Adaptive capacity,	measures to be able to adapt flexibly in the longer term, e.g. temporary buildings;

Climate resilience is created by taking measures that strengthen all of these four capacities. For practical application, the selection of measures is to be made for each specific situation; there is no standard response. Selected solutions will be different each time, depending on the local conditions and on the preferences of the stakeholders.

A climate resilient urban environment can only be achieved by a permanent co-operative effort of all stakeholders including residents. This cooperation must extend not only during the design and construction phases, but throughout the entire life cycle of the system. This means that the planning, design, construction and maintenance processes need an integrated approach. For instance, land subsidence aggravates the vulnerability to flooding and should therefore be stopped in the context of creating a climate-resilient urban environment.

The ambition of creating a climate resilient urban environment is relatively new and as a result examples are scarce. Many are still in the phase of planning and design. To mention a few:

- The introduction of more surface water in new and in existing urban areas to enlarge water storage capacity. In Dutch polder cities the percentage of open water increased from about 4% to about 10% on average.
- In part of a new extension of the city of Gouda called Westergouwe the ground floor level will be raised to 1.4 m above street level to strengthen coping capacity for flooding. A comparable example is found in Japan, in Kanzaki Town in the Saga region (Figure 15.6).
- The Japanese practice of 'uchimizu', which is sprinkling water on the pavement, to reduce ambient air temperatures and the urban heat island, instead of cooling buildings by energy-inefficient air conditioners that raise outdoor air temperature.
- The introduction on large scale of green roofs in Rotterdam to cool the buildings and the urban environment.

In the early eighties, the control of pluvial flooding rather than the need for sustainable development was the reason for experiments with stormwater infiltration facilities and other sustainable urban drainage systems (SUDS) in Japan (Fujita, 1984). Their success has lead to nationwide application of this technology and to a further integration of these facilities in buildings and infrastructural facilities (De Graaf and Matsushita, 2008).

In order to improve the coping capacity for wide-area flooding Japanese government puts an emphasis on effective evacuation schemes and on self-survival in the first 72 hours after a flood disaster. The threshold capacity is strengthened in the Tokyo/Yokohama region with superlevees and a large drainage capacity for their urbanized polder areas. (Stalenberg and Kikumori, 2008; De Graaf and Matsushita, 2008).

15.2.3 Adaptable

Not only climate change but also other long term changes such as urbanization, technological, social and economic developments ask for adaptability and flexibility of the urban area. Rather than building real estate and infrastructure for centuries, we should think of building facilities more temporary in nature, say 30–40 years, that are easily removed after such a period of time. This is necessary to make space for other urban functions and to allow for the use of other design conditions (De Graaf, 2009). Van Nieuwkerk et al. (2010) show that the longevity of the Dutch urban areas hinders climate adaptation, in particular in new towns and urban extensions build after the 70's. Options for temporarily allowing other land use functions are now being studied in order to improve adaptability (Tromp et al., 2010). Ageing population and population shrink rather than urban growth affects numerous settlements, villages and small cities around the world. These demographic developments are another important reason for adaptable urban development. And so are the technological developments in mobility and in information and communication technology.

Figure 15.6 Elevated floor for flood proofing in Saga region, Kanzaki Town, Japan

Not only the buildings and infrastructure, but also the roads, canals, drains and even street furniture of over 30–40 years are ready for revision. Objectives include creating permeable pavements, green roofs, ecological embankments, fishing sites and other facilities that were not yet thought of thirty years ago. Both in the Netherlands and Japan we see large investments in ecological recovery and user-friendly landscaping of urban canals and streams (Figures 15.7 and 15.8). However, the adaptability of buildings and infrastructural facilities is hardly studied yet.

Figure 15.7 Kitazawa, Tokyo Japan: restoration of an old stream in an urban area. Wastewater effluent is used to fill this stream

Figure 15.8 Canal in an urban area in Lelystad, the Netherlands

15.2.4 Healthy

High population densities, a large production of waste and wastewater, urban heat islands and other conditions create a permanent risk for public health in urban areas. Also for urban water managers public health issues are a more or less continuous concern. Where people drink, inhale or touch water there is a potential risk of transferring pathogenic organisms and viruses or health-affecting chemicals. The importance of healthy drinking water goes without saying. Wastewater that enters the public space during pluvial flooding and during combined sewer overflows is full of pathogens and contains high concentrations of pollutants. This pathogenic pollution makes urban surface waters unsuited for bathing (e.g. Schets et al., 2007) – though there are notable exceptions such as the bathing zone of Odaiba in Tokyo. Moreover, the urban surface waters could support development of mosquitoes, vectors for illnesses such as malaria and dengue, if the water system was not properly designed, managed and maintained by the urban water manager.

Until the early 80's of last century public health concerns in urban water management were limited to improving water supply and sanitation systems. Increased environmental awareness triggered a pollution prevention and source control policy that has been in place ever since. Public health protection was widened to *environmental health protection*, to protect ecosystems and biodiversity. The introduction of the European Water Framework Directive in 2000 has put ecological health of surface water at the heart of water quality management. A national monitoring and assessment procedure was defined to evaluate the ecological health of urban canals, ponds and lakes (STOWA, 2001).

At the turn of the century, eco-engineering was added to the spectrum of environmental quality management. 'Building with nature' is used to strengthen the ecological health and robustness of surface waters and groundwater systems. Options for water related ecosystem services - services that contribute to our personal and economical well-being – are developing, e.g. for water-recreational and for water supply purposes (Dionisio et al., 2009). Public health is one of the beneficiaries of eco-engineering; the survival rate of pathogens is minimized by creating adverse conditions in our urban surface waters and in the groundwater.

15.2.5 Pleasant

The perceived quality of the living environment depends on the visibility of water. This was found in several sociological studies (e.g. Van den Berg et al., 2002). Lems sees eleven different values of urban water: biotic, sensitive, analytical, formative, lingual, social, economic, aesthetic, juridical, ethical and pistic. These are the values that make urban water pleasant to have around (Lems, 2008). Strang's anthropological study into the meaning of water for us shows that people assign profound values to water, values that should not be neglected in any plan to change water management (Strang, 2004). Urban planners and urban landscape designers in the Netherlands now learn how to use these values of water in urban design (Hooimeijer et al., 2005) in order to create water cities. Also, in Japan, water is an indispensible element of the urban landscape quality, though less systematically used than in the Netherlands. In Japanese parks and gardens water is a main element for creating a pleasant and relaxed environment (Figure 15.9).

Figure 15.9 Saga castle pond, Japan

15.3 TRANSITIONING TOWARDS WATER CITIES

In order to achieve the 'Water City' and improve mainstreaming of urban water management innovations, many useful approaches have been developed over the last decades in which the connection with urban planning and social amenity is highlighted. Most of these approaches stress the necessity of an integrated system approach that includes the total urban water cycle. An important element in most approaches is the use of an integrated, cross-sectoral, multidisciplinary institutional framework (e.g. Butler and Parkinson, 1997; Niemczynowicz, 1999; Geldof and Stahre, 2006). Governance experts consider the urban water system as a complex adaptive system (Geldof and Stahre, 2006) or socio-technical system (Brown and Clarke, 2007) rather than a technical system.

15.3.1 Mainstreaming of urban water management innovations

As shown in this chapter, a wide range of innovations in urban water management are technically feasible (De Graaf, 2009). These technologies use the urban water system for new functions in order to decrease the vulnerability of urban areas; they reduce the parasitic behaviour of cities and make better use of local resources; they utilise a wide range of water resources instead of only one and reduce the environmental footprint of cities. However, urban water management innovations are often isolated showpieces that hardly contribute to the overall transformation of the urban water system (Brown, 2005: Hunt and Rogers, 2005).

Recent Dutch research (De Graaf, 2009) identified two key conditions for mainstreaming of urban water management innovations. The first condition is including urban water management innovations in spatial planning and development. The second condition is stakeholder receptivity to urban water management innovations. These key conditions will be discussed and will be illustrated with examples from the Netherlands and Japan.

15.3.2 Integration of urban water management in spatial planning

Urban water management innovations will only diffuse to all interventions in the public and private space if they are included in the urban development process. Thus, integration of water management and urban planning is crucial. Transition theory (Rotmans et al., 2001) demonstrates that technologies often have difficulty to get through the acceleration phase. In case the transition succeeds, it takes generations to develop. The process will take place through reinforcing developments in multiple domains, social, cultural economic and technical domains. The multilevel perspective (Rip and Kemp, 1998) shows that most resistance is situated at the level of institutions, technical systems, culture, and legislation (meso level). Mainstreaming urban water management innovations can be achieved by influencing official urban development policy and by evaluation and replication of demonstration projects. Van de Ven et al. (2006) show that this can be done by approaching the innovation challenge as a combination of an optimization problem, a design problem and a negotiation problem.

15.3.2.1 Example from the Netherlands: Rotterdam Water City 2035[1]

The envisioning project Rotterdam Water City 2035 (in Dutch: Rotterdam Waterstad 2035) is an example of successful integration of water management

[1]This section is a summary of: De Graaf, R.E and R. van der Brugge. Transforming water infrastructure by linking water management and urban renewal in Rotterdam, *Technol. Forecast. Soc. Change* (2010), doi:10.1016/j.techfore.2010.03.011

and urban planning. During this project an integral future vision on urban design was combined with a climate adaptation strategy. The objective of the project was to combine the water challenge and the urban challenge. The water challenge was defined as separating clean water and waste water, creating sufficient peak retention capacity, creating seasonal water storage, and creating a denser network of water ways to control ground water. The urban challenge was defined as creating an attractive city for residents and entrepreneurs, stimulating a high social diversity and to strengthen economy vitality. The selected project leader for the Water City 2035 project desired a joint effort of the municipality departments and the waterboards to achieve an integrated approach. Together with the department director, 15 project members were selected: six designers, five water management experts, one member from economic management department and three external members from the waterboards.

The development in the water management regime in Rotterdam shows that urban water management policy is currently integrated with urban planning. Planners of urban water retention infrastructure successfully utilize the opportunities that urban revitalization presents by considering the local context. The role of Rotterdam Water City 2035 has been crucial in changing the perspective. In this policy innovation niche, for the first time urban planners and urban water experts, developed a joint long term vision for the city. Because it was a non-official policy process, more radical ideas and a longer planning horizon were possible than in official policy documents. In this vision, water retention contributes to the upgrading of neighborhoods by increasing living quality. Many innovations that were developed in the project were adopted in official policy. Examples are green roofs, water retention squares and floating urbanization (Figure 15.10). The development of the Rotterdam Water City 2035 policy niche was facilitated by changes in the water management regime during the preceding years. Consequently, urban water professionals were open to the integrated approach that was applied in the project. Adoption of innovative concepts in official policy was further enabled by executive and political support and the presence of change agents in all participating organizations.

15.3.2.2 Example from Japan : Superlevee

The Tokyo lowland consists of river sediments and reclaimed land. The original elevation of this plain is less than two meters above mean sea level. Land subsidence in Tokyo started around 1900 because of exploitation of groundwater. During the post-war reconstruction period groundwater use increased until strict regulatory control was introduced in 1961. In the period to 1965, groundwater use decreased and extraction stopped all together in 1975. By then, some parts of the area has already subsided four meters. This created about 124 km^2 of land that is lower than high tide in Tokyo Bay and 32 km^2 that is lower than low tide. Consequently, levees have been built to protect these lowlands from flooding.

Tokyo is densely populated limiting space to increase the height and width of these traditional levees. A superlevee is a high width river embankment with an urban area that is located on top of the levee. The superlevee can withstand overflows, earthquakes and seepage. In Tokyo, superlevees are located along the Ara River and the Sumida River.

Figure 15.10 The Floating pavilion in Rotterdam is a first step towards self supporting floating districts

Due to the high value of buildings and infrastructure along the rivers in Tokyo and the importance of land ownership, it is not possible to demolish an urban area to construct a levee. Therefore, superlevees can only be created in combination with urban renewal projects. During the planning of an urban renewal project along the river, the river manager is involved and a superlevee is integrated in the urban renewal plan. Characteristic for this type of planning is the long term perspective. One segment of superlevee will have no impact on its own in reducing the flooding probability. This reduction will only be the case if multiple segments of superlevees are created and if these segments are connected in order to form a superlevee riverfront (Figure 15.11). This will take decades to accomplish. The example of the superlevee therefore demonstrates the value of integrating water management and urban planning, and the use of a long term perspective to transform a city in order to make it floodproof.

Figure 15.11 Superlevee in Tokyo, Japan

15.3.3 Stakeholder receptivity

A crucial factor for implementation of urban water management innovation in mainstream practice is the receptivity of professionals and other stakeholders to new approaches, new policies and new technologies. The success of programs that aim to advance the application of a certain policy or technology depends not on the technical quality of these instruments but on the receptivity of the recipients of these change programs. For the full receptivity continuum to be addressed it is necessary that urban water management practitioners (1) are aware of innovations, (2) want to apply these innovations, (3) have the required capabilities, and (4) have sufficient incentives to change their way of working (Jeffrey and Seaton, 2003). The mainstreaming process should be supported by the socio-political context. This includes the active development of values of organisations and regulations.

15.3.3.1 Example from the Netherlands : De Draai Heerhugowaard

A good example of stakeholder receptivity to an urban water management innovation is case study De Draai in Heerhugowaard, the Netherlands. In this case, the municipality was open to using the urban surface water as an energy source.

In the municipality of Heerhugowaard, in the northern part of The Netherlands, is a medium size municipality with 50,000 citizens. In this municipality a new urban district of about 2800 houses, De Draai, will be developed. Similar to many new urban developments in the Netherlands, De Draai has a high planned proportion of surface water. The main considerations to construction with such a high proportion of water are adaptation to climate change and prevention of pluvial flooding. The high proportion of surface water puts the ground exploitation under pressure and results in high costs and loss of land that can no longer be sold to property developers.

A study (De Graaf et al., 2008) demonstrated that the district can be self reliant for heating and cooling purposes by using the urban surface water and urban groundwater for heat collection and cooling (ATES + system). By cooling the surface water with 1.5–1.6°C in three summer months, enough heat is collected to compensate full residential heating and cooling demand. Considerable energy savings and a CO_2 emission reduction of 60% are achieved. Water quality and ecological improvement take place because a lower temperature results in increasing oxygen content. Moreover, the expected water temperature increase by climate change can be partially prevented. The concept is also economically feasible. Considering the full lifetime and all investment and exploitation costs, the concept is more profitable than a conventional system. Therefore, this concept has the potential to contribute to the societal objective to achieve a more sustainable, self-reliant energy supply.

Based on experiences and observations in this project, a number of reasons seem important for the application of this innovation. The ambition of the municipality and their experience with another innovative project, created a window of opportunity for this concept. Also the project team was open to new ideas. The use of the urban water system for energy supply was a potential solution for the financial feasibility of the project. Finally, the proposed water and energy concept won from a competing alternative because of positive water quality impacts.

15.3.3.2 Example from Japan : Water recycling

Subsidy schemes and considering wastewater as a commercial product rather than a waste stimulated the use of recycled water in Japan. Subsidiary systems to enhance recycling of waste water have been introduced and alternative water use has been mandated for large urban development and urban renewal projects. Examples of urban water sources are river water, groundwater, desalinated sea water, stormwater, grey water and treatment plant effluent. By utilisation of the full range of urban water resources in a flexible way, according to the local circumstances, the city is better able to adapt to its physical surroundings. Treated wastewater in Tokyo is used for many urban purposes; there are 560 locations in which recycled water is used. These locations are private buildings, public buildings and large scale recycle schemes (Furumai, 2007). Already in 1979, subsidiary schemes were introduced to promote the use of recycled wastewater system (Matsushita, 2007). Instead of a waste, reclaimed wastewater is considered as a valuable resource with commercial value that is locally available and in constant supply. Approximately 9% of the wastewater is recycled for a variety of purposes (Fujita, 2007). Examples of functions are: road cleaning, sewer cleaning, pavement cooling, train cleaning, urban stream restoration, heating and cooling of buildings, low quality residential purposes, park irrigation, fire fighting, and snow melting on public roads. The Ariake wastewater treatment plant in Tokyo supplies treated effluent to a nearby apartment complex for toilet flushing (Figure 15.12). The wastewater treatment plant supplies the water at a competing rate compared to normal drinking water supply (Van de Ven, 2006).

Figure 15.12 Apartment complex that uses wastewater effluent from the Ariake wastewater plant for toilet flushing in a nearby apartment complex, reducing primary water use

15.4 CONCLUSION

This chapter provided an outline and the features of future water cities. The water city is sustainable, climate robust, adaptable, healthy and pleasant. Water is the key to achieve this. Innovative technologies to this end are now available and have been tested at pilot scale. Mainstreaming them would be the next logical step. Two key conditions for mainstreaming of these urban water management innovations were presented in this chapter. The first condition is including urban water management innovations in spatial planning and development. The second condition is stakeholder receptivity to urban water management innovations. These key conditions were discussed and illustrated with examples from the Netherlands and Japan.

REFERENCES

Adger, W.N., T.P. Hughes, C. Folke, S.R. Carpenter and J. Rockstrom (2005) Social-Ecological Resilience to Coastal Disasters, *Science*, **309**, 1036–1039.

Braungart, M. and W. McDonough (2002) *Cradle to Cradle – remaking the way we make things,/ Cradle to Cradle – afval=voedsel. (2007)*, Sciptum, ISBN 978 90 5594 577 1.

Brown, R.R. (2005) Impediments to Integrated Urban Stormwater Management: The Need for Institutional Reform. Environmental Management, 36(3), 455–468.

Brown, R.R. and J.M. Clarke (2007) *Transition to water sensitive Urban Design: The story of Melbourne Australia.* Report No 07/1 Facility for Advancing Water Biofiltration, Monash University, June 2007, ISBN 978 0 9803428 0 2.

Butler, D. and J. Parkinson (1997) Towards sustainable urban drainage. Water Science and Technology 35 (9): 53–63.

De Graaf, R.E., F.H.M. van de Ven and N.C. van de Giesen (2007) The Closed City as a strategy to reduce vulnerability of urban areas for climate change. *Water Science and Technology* 56(4), 165–173.

De Graaf, R. and J. Matsushita (2008) Stormwater management and multi source water supply in Japan:Innovative approaches to reduce vulnerability. In De Graaf, R. and F. Hooimeijer (eds.) (2008) Urban water in Japan. Urban Water Series 11, Taylor and Francis, London, ISBN 978-0-415-45360-8.

De Graaf, R.E. (2009) Innovations in urban water management to reduce the vulnerability of cities. Feasibility, case studies and governance. PhD thesis, Delft University of Technology.

Dionisio Pires, M., I. Leenen, H.de Man, T. Lycklama a Nijeholt and F. Jol (2009) *Waterrecreatie in de stad; tips en trucs voor het vergroten van de kansrijkheid van projecten.* (Water-recreation in urban areas; tips and tricks for larger opportunities for projects) Report 1200506-000, Deltares, Delft.

Fujita, S. (1984) Experimental sewer system for reduction of urban storm runoff. In: Balmer, P. et al. (eds), *Proc. Third Int. Conf. on Urban Storm Drainage,* June 4–8, 1984, Chalmers University of Technology, Goteborg, Sweden.

Fujita, S. (2007) *Stormwater for urban sustainable water use.* Presentation for Water Research Centre Delft, 15 february 2007.

Furumai, H. (2007) Reclaimed stormwater and wastewater and factors affecting their reuse. In: Cities of the Future: Towards Integrated Sustainable Water and Landscape Management. Editors: Novotny V. and P. Brown.

Geldof, G.D. and P. Stahre (2006) On the road to a new stormwater planning approach: from Model A to Model B. Water Practice & Technology 1 doi: 10.2166/WPT.2006005.

Gemeente Almere (Municipality of Almere) (2008) *De Almere Principles voor een ecologisch, sociaal en economisch duurzame toekomst van Almere 2030.* Gemeente Almere.

Hooimeijer, F., H. Meyer, A. Nienhuis (eds) (2005) *Atlas of Dutch water cities*. Sun Publishers, Amsterdam, ISBN 90 5875 184 8.

Hunt, D.V.L. and Rogers C.D.F. (2005) Barriers to sustainable infrastructure in urban regeneration, engineering sustainability. Proceedings of the Institution of Civil Engineers, 158, (ES2), 67–81.

Jeffrey, P. and R.A.F. Seaton (2003) A conceptual model of 'receptivity' applied to the design and deployment of waterpolicy mechanisms. Environmental Sciences 1, 277–300.

Landelijke Projectgroep Watertoets (2003) Handreiking Watertoets 2; Samewerken aan water in ruimtelijke plannen. V&W RIZA, Den Haag/Lelystad.

Lems, P., (2008) *The values of water*. In: Proceedings 11th International Conference on Urban Drainage, Edinburgh, Scotland, UK.

Lloyd, S.D., T.H.F. Wong and C.J. Chesterfield (2002) *Water Sensitive Urban Design: A Stormwater Management Perspective*. 02/10 Industry report, CRC for Catchment Hydrology, Australia.

Mouritz, M. (2000) Where to now? – Keynote paper. *First National Water Sensitive Urban Design Conference*, 30th-31st August 2000, Melbourne, Australia.

Niemczynowicz, J. (1999) Urban hydrology and water management- present and future challenges. Urban water, 1, 1–14.

Rip, A. and R. Kemp (1998) Technological Change. Human Choice and Climate Change, in: S. Rayner, and E.L. Malone. Volume 2: 327–399, Battelle Press, Columbus, Ohio, USA.

Rijsberman, M.A. and F.H.M. Van de Ven (2000) *Different approaches to assessment of design and management of sustainable urban water systems*. Environmental Impact Assessment Review (20), pp. 333–245.

Rotmans, J., R. Kemp and M. Van Asselt (2001) Transition Management in Public Policy, Foresight 3, 15–31.

Schets, F.M, J.H. van Wijnen, H. Schoon et al. (2007) *De microbiologische kwaliteit van het grachtenwater in Amsterdam*. RIVM rapport 330000011/2007, Bilthoven.

STOWA (2001) *Ecologisch beoordelingssysteem voor stadswateren (Ecological evaluation system for urban water)* Nr 2001-17, STOWA, Utrecht, ISBN 90-5773-130-4.

Stalenberg, B. and Y. Kikumori (2008) Urban flood control on the rivers of Tokyo metropolitan. In De Graaf, R. and F. Hooimeijer (eds.) (2008) Urban water in Japan. Urban Water Series 11, Taylor and Francis, London, ISBN 978-0-415-45360-8.

Strang, V. (2004) *The meaning of water*. Berg, ISBN, 1 85973 748 X.

Team Urban Water Management (2009) *Land & water management in the urban environment*, Deltares, Delft.

Tromp, E., Franssen R. Veen, van der, B (2010) *Tijdelijke kennisagenda Tab: bouwsteen voor Tab Innovatieprogramma 2010-2011*.Projectnummer 1200244. Deltares. Utrecht.

UNFPA (2007) *State of World Population 2007*, ISBN 978-0-89714-807-8, United Nationas Population Fund, New York, USA.

Van Beek, V.M. and M.W.A. Hounjet (2009) GeoCheck; beheerst bouwen op een slappe ondergrond. From: http://www.deltares.nl/nl/product/151715/geocheck/ 187445 (visited Aug 4, 2010).

Van de Ven, F.H.M, E. Luijendijk, M. de Gunst, E. Tromp et al. (2009) *Waterrobuust Bouwen; de kracht van kwetsbaarheid in een duurzaam ontwerp*. Beter Bouwen Beter Wonen / CUR Bouw & Infra, Gouda. ISBN 978 90 376 0513 6.

Van de Ven, F.H.M., S.P. Tjallingii, P. Baan, P.J. Van Eijk and M.A. Rijsberman (2006) Improving urban water management planning. In: Proceedings of the International Symposium on Lowland Technology, S Hayashi, H Araki & K Hokao (Eds.), Saga, Japan, pp. 573–578.

Van den Berg, A., M. Bloemen, T. de Boer and J. Roos-Klein Lankhorst (2002) *De beleving van watertypen: Literatuuroverzicht en validatie van de indicator "water" uit het belevingsGIS*. DLO Natuurplanbureau-onderzoek. Werkdocument 2002/xx, Alterra, Wageningen.

Van Dijk, J.M. (2008) *Water and Environment in Decision Making. Water assessment, Environmental Impact Assessment and Strategic Environmental Assessment in Dutch Planning. A comparison.* PhD thesis Wageningen University. ISBN 978-90-8504-877-0.

Van Nieuwkerk, E., J. Rijke, W. Veerbeek, K. Stone, C. Zevenbergen and F. van de Ven (2010, in prep.) *Climate resilience of urban areas in the Netherlands.* PBL -Netherlands Environmental Assessment Agency, Bilthoven.

Wong, T.H.F. and M.L. Eadie (2000) Water sensitive urban design – a paradigm shift in urban design. *Proceedings of the International Water Resources Association for the Xth World Water Congress*, 12th–16th March 2000, Melbourne, Australia.

Chapter 16

System solutions in urban water management: the Lodz (Poland), perspective

Iwona Wagner and Michael Zalewski

University of Lodz, Department of Applied Ecology and International Institute of Polish Academy of Sciences – European Regional Centre for Ecohydrology under the auspices of UNESCO Lodz, Poland

ABSTRACT

Quality of life and health in cities is becoming more and more important but challenged due to increasing human population, urban sprawl and climate change. New ways of planning and managing for sustainable cities are needed. Ecological components of urban systems should be recognised as essential regulatory drivers, critical for enhancing the capacity of ecosystems against human impact, providing ecosystem services and cost-efficient solutions to the cities' issues. Understanding the interrelationship between hydrological and biological processes can provide, according to ecohydrology theory, the methods and tools to do so.

Such an approach has been tested in Lodz, Poland (Lodz is one of the demonstration cities of the EU Project SWITCH (GOCE 018530), and a Global Reference Site on Ecohydrology for UNESCO IHP (International Hydrological Programme)). Transdisciplinary research provided a basis for developing and demonstrating ecohydrological system solutions. Creation of the Lodz Learning Alliance empowered implementation and upscaling of innovative solutions. As a result, a new concept to incorporate water (blue) and green areas as a template and an integral part of the city planning ('blue-green network') was formulated, which paves the way for sustainable development in the city and adaptation to global climate changes.

16.1 INTRODUCTION

The predicted sharp increase of population in urban environments over the coming decades will amplify existing problems related to declining resource availability, efficiency of transportation, recreation, quality of life, and human health. These problems can be further exacerbated by urban sprawl and the effects of climate change. Quality of life and health in cities will become more and more critical. New ways of planning for sustainable 'low emission and friendly urban landscapes' that encourage city dwellers to adopt attractive and healthy life styles, are needed. Water issues will in many cases play a central role in this process. Affordable, flexible technologies attuned to environmental conditions and settings; efficient and reliable water infrastructure and facilities; appropriate water quality for different uses; and capable and adaptable integrated urban water management (IUWM) at affordable cost, will be indispensable ways to get closer to sustainability. One of most important challenges nowadays is to recognize that management of water in cities goes beyond drinking water provision and sanitation. Water and related ecosystem services play a fundamental role in creating quality of life and sustainable development. These approaches are not easy to develop and introduce in the context of rigid and determined engineered and institutional structures of existing cities. There is a need for holistic city management which recognizes the importance and role of the ecological component of urban systems as one of their essential regulatory drivers (Zalewski & Wagner 2005; Wagner & Zalewski, 2009). Understanding and managing highly altered ecological processes driving the 'novel ecosystems' (*sensu* Hobbs et al., 2006) of urban space is critical for reducing environmental degradation and for providing cost-efficient solutions for water management. Water and the urban environments also need to be functionally integrated with the city architecture and spatial distribution.

One of the ways to improve the scope of ecosystem services in highly impacted urban systems is to understand the interrelations between hydrological and biological processes which, according to ecohydrology theory (Zalewski et al., 1997; Zalewski, 2002), provide a method and tools to enhance the capacity of ecosystems to absorb human impact. Application of this approach, complementing traditional engineering solutions, has been proposed and tested in Lodz, Poland. An important result was the development of a new concept to incorporate water (blue) and green areas as a template and an integral part of the city planning process. This 'blue-green network' paves the way for sustainable development in the city and adaptation to global climate changes.

Project activities towards developing a paradigm shift in Lodz occurred at three levels:

- transdisciplinary research as the basis for developing system solutions, ecohydrology application at a range of scales, and
- creation of the Lodz Learning Alliance and strengthening of institutional processes.

16.2 LODZ–CITY ON THE 'EDGE'

Lodz, with 800,000 inhabitants, is the second biggest city in Poland, located in the central part of the country. In the 19th century, the water resources of Lodz became one of the incentives for its development, transforming it from a provincial town to a major European manufacturing centre. Most of the watercourses were used to provide power for the developing textile industry. Over time these watercourses were channelized and gradually converted into a sewerage system. This process, together with dense development of the city, led to a fundamental degradation of the water cycle. This had adverse consequences for Lodz's urban areas, such as reduced infiltration and ability of the landscape to retain water, decreased evapotranspiration and intensified runoff during storm events. With time, these issues started to hamper the quality of life and raise the cost of water infrastructure that was designed using what was then an accepted traditional engineering approach.

The 20th century transformations and depressions in the East European economy decimated markets for the city's textiles and set new challenges for its development. Searching for new opportunities to bring the city forward drew attention to the importance of creating a healthier urban environment, better living spaces and improved quality of life as means to attract new investment and activities. Little by little it became evident, that the city's water resources would once again become key for its more sustainable development. The city started to look at the possible restoration of its rivers as a mean to enable closer association with green spaces and the revitalization of the historical industrial architecture of the city.

Given the specifics of the Lodz location, rehabilitating its ecological conditions could also contribute considerably to its ability to adapt to global climate change. Lodz is situated on the 'edge' between the watersheds of two major rivers in Poland, the Vistula and Oder. The 18 streams (average flow < 1 m3 s-1) that have their sources in the city and its surroundings are now part of a combined sewerage system, literally draining Lodz by conveying rainwater quickly towards the outskirts of the City. This drying process may deepen due to a hydrological drought that is projected as a result of global climate change in this part of the Poland. Lodz's water management strategies can either remain unchanged, reinforcing the existing environmental problems and increasing overall costs for the city management, or they can be open to new management options and contribute to its sustainable development.

A switch was needed in the perception of stormwater–from a threat and waste to be removed quickly from the city area (as in traditional management of urban water resources) to a valuable resource, essential to be retained in the urban space for restoring the water cycle in the landscape. Social and economic benefits resulting from this changed approach can help in adapting to global climate change including among others: reduced costs of development of stormwater infrastructure for managing storm events, reduced costs of flood management and economical losses by society, improved humidity and air quality under increasing drought conditions, providing new attractive recreational spaces for the city dwellers (Figure 16.1 a, b).

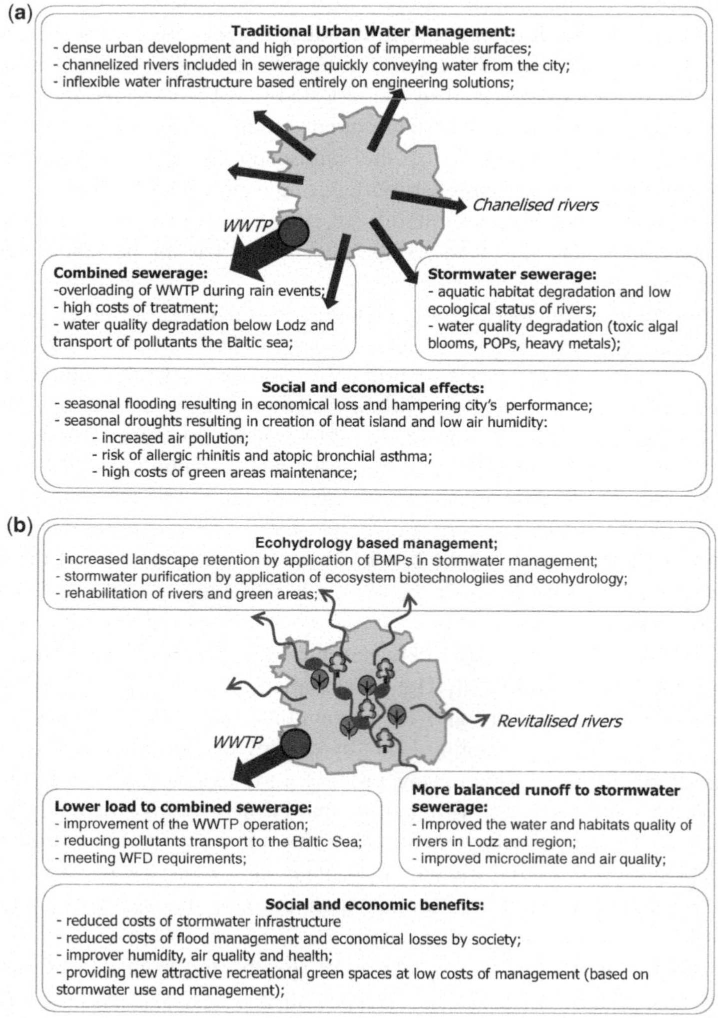

Figure 16.1. Ecohydrology as a basis for a systems approach to urban stormwater management: change from (a) an over-engineered system accelerating stormwater transfer out of Lodz to (b) a system retaining and slowing down water transfer from the city by enhancement of interception and infiltration in green areas, storm water purification and retention.

16.2.1 Ecohydrology–a more sustainable way forward

The interdependence between hydrological and biological processes and their fundamental importance for ecosystem functioning has been described in literature for decades (e.g. Eagelson 1982; Zalewski & Naiman 1985; Naiman, et al., 1988; Baird & Wilby 1999, Rodriguez-Iturbe 2000), as well as quantified and validated at various scales, such as landscape (e.g. Kędziora et al., 1989; Baird & Willby 1999; Kędziora & Olejnik 2002; Ryszkowski 2002), floodplains and wetlands (e.g. Junk et al., 1989; Agostinho et al., 2004; Mitsch et al., 2005, 2008), rivers (e.g. Zalewski et al., 1985), lakes and reservoirs (e.g. Ploskey 1985; Zalewski et al., 1990; Tundisi & Straskraba 1999) or coastal zones (e.g. Chicharo et al., 2001; Wolanski et al., 2004). Interconnections between hydrological and ecological processes in the environment underpin the ecohydrology concept as a holistic framework, which integrates this knowledge in order to enhance catchment resilience against human impact, contribute to water and environmental sustainability and strengthen the provision of ecosystem services (Zalewski et al., 1997). Ecohydrology, as a problem-solving, transdisciplinary science, strives to achieve this by considering interrelations between the catchment (e.g. as a template for water and nutrient dynamics) on the one side, and biological processes on the other (e.g. from ecological succession, habitat modification, biological productivity down to nutrients circulation by the microbial loop), and using them as a tool for 'dual regulation' (i.e. regulating biological processes by controlling hydrological parameters of an ecosystem and the other way around – regulating hydrological processes by managing biota) (Zalewski 2006).

16.2.2 Applying the ecohydrology to urban areas

Ecohydrological knowledge, expertise and know-how has been generated for 15 years now, mainly in semi-natural, large-scale systems (e.g. Zalewski, Tarczyńska, Wagner-Lotkowska 2000; Chicharo et al., 2001; Agostinho et al., 2004; Wolanski et al., 2004; Wagner et al., 2008). Although the scale and opportunities for its implementation in catchments significantly differs from those in cities, much of the original knowledge has been recently translated into urban systems (Zalewski & Wagner 2005; Wagner & Zalewski 2009, Wagner et al., 2008). The philosophy of enhancing the absorption capacity of ecosystems against human impact, within limited urban space and under high anthropogenic pressures, provides a new challenge for ecohydrology (Wagner et al., 2008). It sets new goals for ecohydrological management, to include improvements in the quality of life and increases in the efficiency of environmental management, while reducing the costs of investments in infrastructure.

Implementation of ecohydrology is based on a four step methodology (Wagner et al., 2009), including i) monitoring of threats (Tarczynska & Zalewski, 1995),

ii) analysis of the cause-effect relationships (Zalewski et al., 2000), iii) development of methods (Kiedrzynska et al., 2008), and iv) elaboration of system solutions (Zalewski, 2000).

Activities taking place in Lodz followed the above approach through three types of action: a) undertaking transdisciplinary science in developing system solutions, b) implementing results of the research at a range of scales, and c) strengthening institutional processes by establishment and development of a Learning Alliance for scaling up implementation and generating a strategy for further development.

Undertaking transdisciplinary science for development of system solutions: Most of the research was focused on understanding of ecohydrological processes driving urban water systems. Within the river systems, it addressed such aspects as seasonal and spatial analysis of hydrological pattern of rivers, water quality, ecosystem dynamics including phytoplankton community with special emphasis on toxic cyanobacterial blooms, zooplankton communities and fish assemblages, among others. One river valley was also researched with the focus on vegetation mapping and landscape validation. In parallel, research on the application of sewage sludge for biomass production and removal of heavy metals by phytotechnology[1] (UNEP, 2003) was conducted at the wastewater treatment plant (WWTP) (Drobniewska, 2008). Following the idea of the demand driven research (Gyawali et al., 2006), the results of the ecohydrological research provided the basis for the development of systemic solutions for the city, and for decision making in river restoration, reservoir construction, stormwater management and in managing the conversion of sewage sludge into bioenergy (Table 16.1).

The ecohydrological approach in urban systems has substantially strengthened links between classical ecohydrology and such aspects as institutional analysis, social research, economy, health and engineering. Stakeholder analysis allowed for building a comprehensive knowledge base on the institutional background for water resources management and decision-making processes in the city, and allowed for the efficient involvement of the key stakeholders in steps towards the paradigm change. Research on perceptions and attitudes towards the key socio-economic drivers of water quality by the key stakeholders allowed prioritisation of actions for the Lodz water strategy. Sociological research has also allowed analysis of local community needs, perceptions, visions and attitudes to current and proposed land uses within the planned city parks and to river restoration and stormwater management (Krauze & Wlordarczyk, 2010).

System solutions, linking ecohydrology-based management methods with social and economic challenges and opportunities, contribute not only to a better environmental quality of socio-ecological systems, but also to the overall wellbeing of a region, its ecological security and sustainability (Wagner et al., 2008). The idea of a knowledge-based society and economy implies providing for

[1]The term phytotechnology describes the application of science and engineering to examine problems and provide solutions involving plants.

Table 16.1. The key research areas for development of the Blue-Green Network concept and its implementation in the Lodz City

Research	Research application
Stakeholder analysis;	Building a comprehensive knowledge base on the institutional background for water resources management in the city, identifying gaps and overlaps in competences, identifying mechanisms for decision-making processes.
Defining key socioeconomic drivers of water quality in the city and elaboration of their risk profiles; (Krauze & Wlodarczyk, 2010)	Building a comprehensive knowledge base on the drivers of water resources; understanding people's attitudes behind the decision-making process, and elaborating priority actions for Lodz water strategy
Seasonal and spatial analysis of water quality, persistent organic pollutants (POPs) and cyanobacterial toxins in reservoirs;(Wagner et al., 2007; Urbaniak et al., 2008; 2009)	Assessment of risk for safe recreational use of reservoirs;Improved design of stormwater reservoirs for increased absorption capacity against pollution;
Hydrological monitoring on line;	Developing water budget and mathematical models for stormwater management in Lodz;
Study of ecohydrological relationships in rivers; (Wagner et al., 2007)	Design of sedimentary-biofiltration system for stormwater purification;
Vegetation mapping and landscape validation of the Sokołowka River valley; Analysis of current and proposed land uses within the Sokolowka River Park against local community needs, visions and attitudes	Elaboration of river valley rehabilitation plan; Creation of a project for Sokołowka River Park;
Assessment of the effect of aquatic ecosystem contamination on the condition of fish at molecular level (RNA/DNA ratio); (Mankiewicz-Boczek et al., 2010)	Possible indicator for biomonitoring urban rivers degradation and a screening method for evaluation of contamination of water ecosystems;
Research on the application of sewage sludge for biomass production; (Drobniewska, 2008)	Increasing the yield of bioenergy from willow crops and alternative utilisation of sewage sludge;
Research on river restoration, landscape validation and stormwater management; Analysis of the role of urban agriculture for quality of water, biodiversity and cultural values of landscape	Development of a 'Blue-Green Network Concept'

long-term development, increasingly innovative research and economic measures and contributing to removing barriers and social risks, such as unemployment and social exclusion (Buchner-Jeziorska, 2010). Key research activities contributing to system solutions in Lodz are presented in Table 16.1.

Implementing results of the research at a range of scales – the water management challenges in Lodz have been initially addressed by two demonstration projects (Wagner & Zalewski 2009; Box 16.1, 16.2). The demonstration projects have an important role in making SWITCH visible in the city. They have proven to be an effective way to show new solutions and are useful test cases for cooperation between different stakeholders. In this regard, they are also key elements of the strategy to scale-up new implementation, education and enhanced dissemination (Wagner et al., 2010). Another strategy to continue and expand the work done in SWITCH and put the research into practice includes scaling up implementation of the technologies which were tested in the demonstration projects and maximising impact on a wider scale, by securing funds from the EU and other sources. The new projects build on the

Box 16.1 Demonstration: Restoration of the Sokolowka River

The Sokolowka River (average flow: 0.17 m3 s-1) flows across the northern part of the city and is supplied with about 50 stormwater outlets. It was channelised by concrete slabs which straightened the course and deepened the bed. However the middle section of the river valley has maintained its semi-natural character with patches of meadows, wetlands and forests.

Demonstration activities harmonized the existing hydrotechnical infrastructure with ecohydrological measures and included: i) river valley and river channel rehabilitation for improving ecological status of a river, increasing water retentiveness in the landscape and groundwater level, and vegetation support; ii) mitigation of extreme stormwater flow by the construction of cascade of reservoir; iii) stormwater purification in Sedimentary-Biofiltration System and by application of phytotechnology; iv) creating space for education and rising environmental awareness.

Implementation of demonstration activities and awareness raising has triggered several bottom–up initiatives including developers implementing best management practices (BMPs) in stormwater management in their new investments (Figure 16.2).

Demonstration activities on the Sokolowka river aim to increase the quality of life and aesthetic values in the area and to contribute to an improved, healthier environment for city dwellers. They have become an inspiration for the development of a framework concept for other rivers in Lodz, and for the formulation of a 'Blue-Green Network' concept. Experiences from this project are now being implemented on another river in Lodz in the framework of the EU Life+project[2].

[2]EHREK: Ecohydrologic rehabilitation of recreational reservoirs 'Arturówek' (Lodz) as a model approach to rehabilitation of urban reservoirs (LIFE PLUS, LIFE08 ENV/PL/000517).

research, focus on implementation and have a strong component of communication and dissemination.

The demonstration projects became an inspiration for several developers, architects, artists, spatial planners, NGOs, scientists and others, and initiated bottom-up activities that contributed to spreading the SWITCH expertise and

Box 16.2 Demonstration: Bioenergy – Using sewage sludge as a fertilizer in willow plantations

This demonstration project was located within the protected zone of the wastewater treatment plant, on the Ner River). The Ner (catchment area: 1865 km2; mean discharge: 0.8 m3 s-1; length: 126 km) is a tributary of the Warta River. The river receives treated wastewater from Lodz and its satellite cities. The WWTP processes about 210,000 m3 of wastewater per day, retaining about of 215 tons of sewage, sands and sewage screenings, and producing about 70,000 tons of sewage sludge per year stored in the lagoons. The sludge contains high concentrations of nutrients that makes it a valuable resource, e.g. as fertilizer for short rotation forestry, such as an energy producing willow plantation (the willow is used as a fuel to replace fossil fuels). A major problem is the high level of heavy metal pollutants in the sludge, which restricts the sludge use as a fertilizer. Lowering the heavy metals concentration is not easy to resolve because it requires action by the city authorities to control the pollution at source (controlling the companies that are likely to be involved in pollution).

The company operating the plant – an independent government-owned enterprise and a Lodz learning alliance member – has been closely involved in the research and the demonstration. Pilot plantations of willow on 6.2 ha of the total site area of about 60 ha were established in 2004 and cultivated by the company on municipal land in the protected zone next to their wastewater treatment plant.

The demonstration project addressed the following issues: i) estimation of the efficiency of closing nutrient and anthropogenic cycles by applying sewage sludge as fertilizer on experimental willow plantations to be used as a source of energy; ii) application of phytoremediation for inactivation and removal of heavy metals present in the sewage sludge; iii) estimation of the potential for enhanced biomass production and bioenergy generation; and iv) developing mathematical model for the decision support system for the energy-producing plantation operation.

Extending the sewage treatment by willow plantations and constructed wetlands can generate several benefits (Figure 16.3). Improvements in water quality increase the quality of life in the city. Willow plantations, based on a mixture of selected natural and fast-growing species, not only maintain biodiversity but can also provide an alternative source of energy (bioenergy) that can help to reduce CO_2 emissions from burning fossil fuels. The resultant ash can be used to fertilise forest plantations. Production of bioenergy generates new employment opportunities and revenue flows while reducing capital outflows for fossil fuel use.

The WWTP is now producing biomass which is being used by neighbouring counties (the smallest administrative unit in Poland) for heating schools and public buildings.

prompting ideas for implementation of innovative approach around the city. Finally, they also contributed to development of the city-scale concept of a Blue-Green Network.

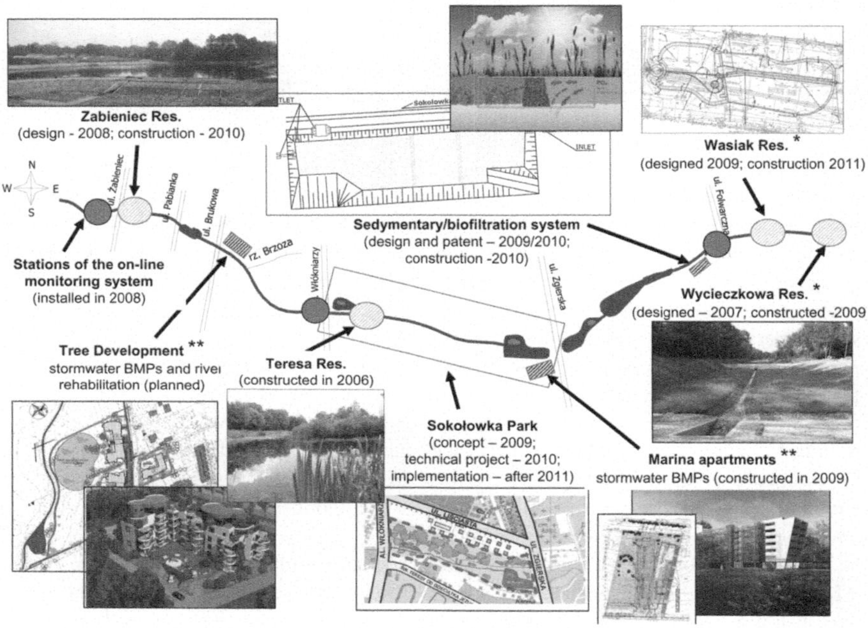

Figure 16.2. Interventions linked to the implementation of ecohydrology in urban space within the SWITCH project in the demonstration Project on the Sokolowka River: constructed, under development and designed, SWITCH spin-offs and constructions developed using the SWITCH generated knowledge (*), bottom-up initiatives resulting from the Learning Alliance activities (**) (Wagner & Zalewski, 2009; updated)

Learning alliance and institutional process - Innovation is increasingly needed to identify key issues in a city, develop a research process to quantify them, and to communicate and implement the results that will resolve issues and problems. The underlying rationale is to improve the impact of research on policy and outcomes while taking into account the prevailing institutional context. A change of a paradigm in water resources management is only possible based on good co-operation between all the city stakeholders at the same level of participation, and by assuring good communication and mutual learning. Such opportunities were created through establishing the city learning alliance (Butterworth et al., 2008) in Lodz in 2006. The group was initially formed by a few key stakeholders, including representation from the City of Lodz Office (the city executive) and the key water services provider. The membership has been constantly growing with time, and by the end of the project the alliance amounted to more than 60 people and included a broad representation from the city stakeholders. Strong new linkages have been forged between scientists, decision makers, water service providers, media NGOs and other stakeholders creating the potential to scale up the use of research results. Members from national, regional and local levels (Figure 16.4) started to influence other projects in the city, addressing water issues, sustainable development and spatial planning, restoration and investments.

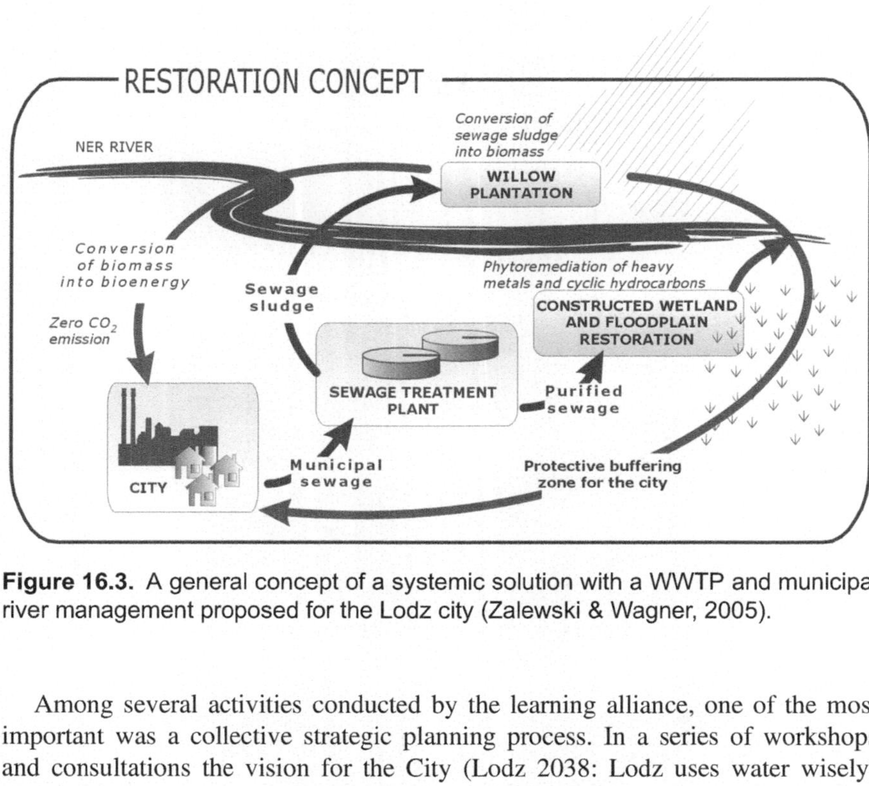

Figure 16.3. A general concept of a systemic solution with a WWTP and municipal river management proposed for the Lodz city (Zalewski & Wagner, 2005).

Among several activities conducted by the learning alliance, one of the most important was a collective strategic planning process. In a series of workshops and consultations the vision for the City (Lodz 2038: Lodz uses water wisely) was developed, followed by the outlining of four possible future scenarios and then a number of strategic options for achieving the vision. Short- and long-term strategies, goals, indicators and phases for activities based on these documents are now being formulated. As a result of this process the learning alliance provided recommendations for one of the city's most important strategic documents 'Study of Conditions and Directions of Spatial Development of the City of Lodz,' and embedded the SWITCH recommendations into other strategic documents of the city.

Thanks to the broad co-operation achieved with city decision makers, some aspects identified in the vision have already been followed up. Positive steps, were taken while production of a final strategic document was still in progress, include implementation of best practice in stormwater management in new developments by both government and private investors, establishment of new green areas, fundraising for other river rehabilitation schemes and increasing the overall awareness about ecological approaches.

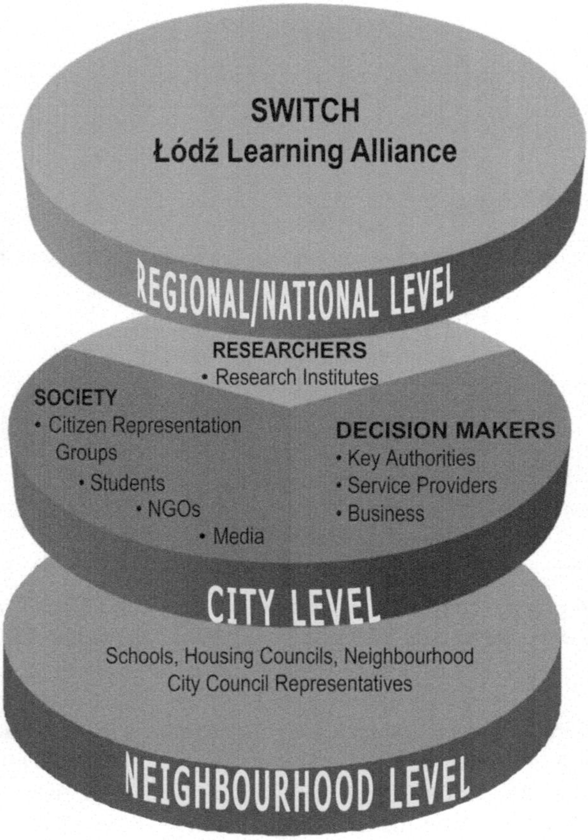

Figure 16.4. Structure and memberships in the Lodz learning alliance

16.2.3 Blue green network

Originally, SWITCH only planned to work in Lodz at a demonstration scale. However, learning alliance interest resulted in the project being expanded to activities at city scale, including recommendations for the city's sustainability planning based on water resources as a key element. This has proved vital to the uptake of integrated urban water management at a larger scale.

In 2008, the concept of a Blue-Green Network was developed as a result of the SWITCH project by the ERCE UNESCO team (Zalewski & Wagner 2010, in press). The concept links development of a network of (possibly) restored rivers and green areas (agricultural areas, parks, old orchards, wastelands, degraded areas and others) as a basis for spatial planning, sustainable development and adaptation to global climate change.

The Blue-Green Network (Figure 16.5) is a concept of urban space development for the city of Lodz, that complements the traditional city planning concept of a green belt around the city. Spatial organization proposed by the concept is based on a network of existing rivers and their corridors, parks and forests, linked to each other and to new proposed municipal parks. Buffer zones around the green areas and rivers, will contribute to the protection and maintenance of the continuum of ecological processes within the network.

In the new challenges that Lodz is currently facing, the concept provides new perspectives for reducing threats and strengthening opportunities for sustainable development and adaptation to global climate changes. The new public space will contribute to several social and environmental functions, such as:

- increasing the number of areas for recreation and psychophysical regeneration, with better access from each section of the city, contributing to promotion of healthy life styles and better quality of life;
- better conditions for developing zero-CO_2-emission alternatives for individual and public transportation by providing space for bike routes, pedestrians and trams;
- providing areas for stormwater management through ecohydrological measures and ecosystem biotechnologies, such as application of biofiltering systems, constructed wetlands, river rehabilitation, building reservoirs with increased capacity due to phytotechnology applications, and other measures. Increased retention of purified stormwater in the city landscape will result in lower runoffs during stormwater events, lowering the costs of investments in stormwater infrastructure, reducing economic losses after flooding, and improving functioning of the WWTP and the city rivers. This approach constitutes a flexible and cost-efficient adaptation to global climate change;
- improving humidity, air quality and microclimate in the city to contribute to a healthier environment and to reduce the number of cases of asthma and allergies;
- protection and revitalisation of the historical, environmental and cultural heritage of the city, contributing to improvements in its attractiveness, aesthetics and quality of life.

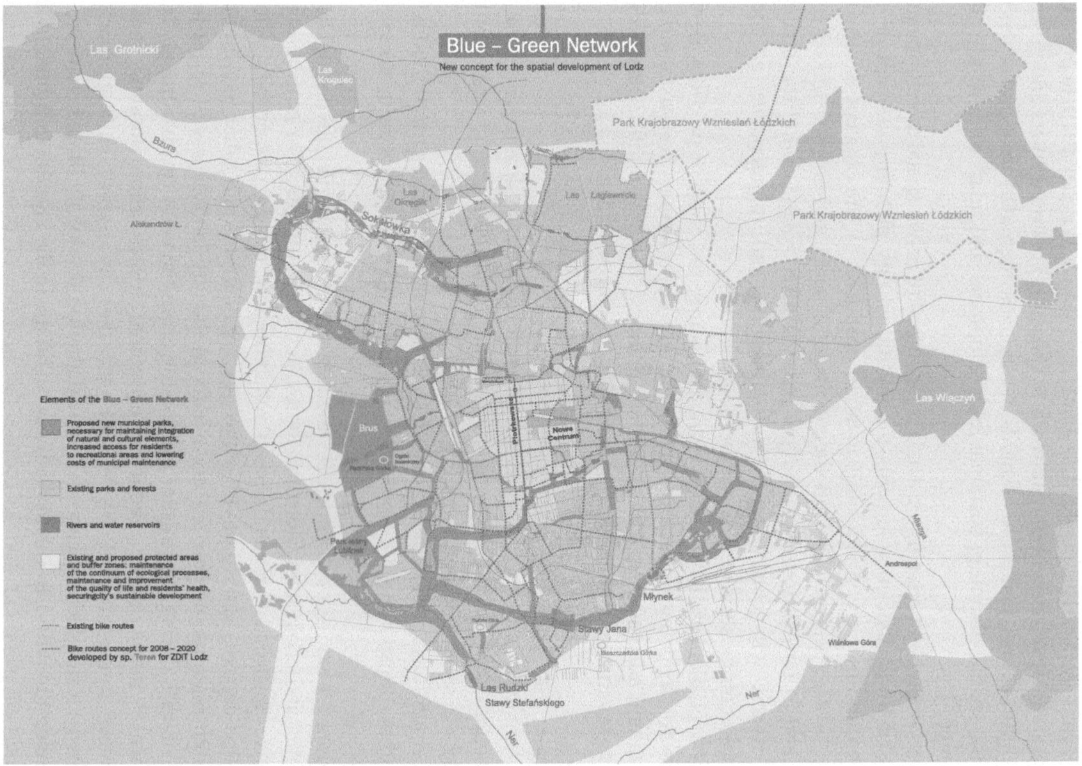

Figure 16.5. Blue-Green Network in Lodz City (dark green - proposed new municipal parks; light green - existing parks and forests; blue - rivers and water reservoirs; yellow - existing and proposed protected areas and buffer zones; red dotted line - existing bike routes; violet dotted line–proposed bike routes)

In February 2009 the concept was presented at a conference organised by the Mayor of Lodz. The city management and stakeholders thought it a promising idea for sustainable development of the city. Information about the Blue-Green Network was disseminated to local media and has been presented in various meetings and local, regional and international events in 2008–2010, including the World Water Week in Stockholm and the International Water Week in Singapore. In Lodz, several meetings were held with a wide range of stakeholders and the idea continued to capture local interest. An effort was also made to reach politicians through meetings with three City Council committees and other stakeholders participating in the spatial planning of the city.

The Blue-Green Network emerged during the course of the project as a framework that extended the initial scope of SWITCH to wider city development processes. Through this framework, water was put on the city agenda and innovations developed in SWITCH will be more widely applied in the future.

16.3 CONCLUSIONS

1. **Blue-green network:** There is now a wide acceptance that water and green areas have an important role to play in the future of Lodz. Direct involvement of multiple municipal departments has been an important feature in uptake of SWITCH concepts. Interest in 'green and blue' spaces in the city has increased and new initiatives have developed as spin-offs.
2. **Paradigm change:** Paradigm change and implementation of case-specific innovations in urban water management requires understanding and consideration of local community needs, visions and attitudes, as well as results of demand-driven, integrative scientific research, based on understanding of local opportunities and limitations.
3. **Demonstration projects:** Demonstration projects have proved to be an effective way in making research and innovations visible in the city. They trigger bottom-up initiatives and are key elements of the strategy to scale-up new implementations and enhance dissemination. They are also useful test cases for cooperation between different stakeholders.
4. **Learning alliance and institutional co-operation:** The learning alliance proved to be most useful in information sharing, getting a better understanding of the needs, priorities, limitations and perspectives of different stakeholders and improving coordination among them. It also facilitated the development of a consensus about the way forward and the formulation of the first Lodz vision for water resources and their integrated management in the city of the future.
5. **The way forward:** SWITCH has catalysed in Lodz a process of paradigm change that is likely to continue after the end of the project. Several activities have been already undertaken to ensure continuation of research, demonstrations, training and dissemination, in order to further consolidate

achievements and collaborative thinking to date. Continuation of these efforts is essential to ensure a more sustainable management and development of the city.

ACKNOWLEDGEMENTS

Members of the research team of SWITCH Lodz have contributed to many aspects of the project. Unless indicated otherwise, all the researchers are working in the Department of Applied Ecology, the University of Lodz or the International Institute of Polish Academy of Sciences – European Regional Centre for Ecohydrology under the auspices of UNESCO:

- **Biological and toxicological assessment and monitoring**: Jonna Mankiewicz-Boczek, Katarzyna Izydorczyk, Joanna Zelazna-Wieczorek (Department of Algology and Mycology University of Lodz), Agnieszka Bednarek, Adrianna Wojtal-Frankiewicz, Tomasz Jurczak, Zbyszek Kaczkowski, Magdalena Urbaniak, Maja Godowska, Ilona Gagała, Marek Ubraniak;
- **Hydrological processes**: Wojciech Frątczak; Marek Zawilski and Grazyna Sakson (Department of Environmental Engineering Subdepartment of Water and Sewage Technology, Technical University of Lodz); Adam Bartnik and Przemysław Tomalski (Department of Hydrology and Water Resources, University of Lodz);
- **Bioenergy – conversion of sewage sludge into bioenergy**: Agata Drobniewska;
- **Landscape and river valley vegetation**: Kinga Krauze, Edyta Kiedrzynska; Józef Kurowski, Marcin Kiedrzynski, Piotr Witoslawski (Department of Geobotanic and Plant Ecology, University of Lodz);
- **Social aspects**: Kinga Krauze, Renata Wlodarczyk;
- **Human Health**: Piotr Kuna, Izabela Kupryś-Lipińska (Department of Internal Medicine, Asthma and Allergy, Norbert Barlicki Memorial Teaching Hospital No. 1, Medical University of Lodz); Nofer Institute of Occupational Medicine in Lodz;
- **Learning Alliance**: Monika Dzięgielewska-Geitz, Łukasz Kamiński,

The City of Lodz is a proposed platform for Long-Term Socio-Ecological Research (LTSER) as part of the European Long-Term Ecosystem Research Network.

REFERENCES

Agostinho, A.A. Gomes, L.C. Verissimo, S. & Okada, E.K., 2004. Flood regime, dam regulation and fish in the Upper Paraná River: effects on assemblage attributes, reproduction and recruitment. Reviews in *Fish Biology and Fisheries* 14: 11–19.

Baird, A.J. & Wilby, R.L. eds., 1999. *Eco-hydrology. Plants and water in terrestrial and aquatic environments.* Routledge, London, New York.

Buchner-Jeziorska A. (2010) 'Globalised Knowledge-based Society and New Social Risks and Universities. The Polish Study', w: European Studies on Inequalities and Social Cohesion nr 1, Łódź: Uniwersity Press; s. 7–40.

Butterworth, J. A., Dziegielewska-Geitz, M., Wagner, I., Sutherland, A., Manning, N., Da Silva, C. & Verhagen J., 2008. *Learning alliances for innovation in urban water management.* Paper presented at the thematic workshop Water and Cities workshop, Water Tribune Expo Zaragoza, 25-28 July 2008.

Chicharo, L., Chícharo, M.A., Esteves, E., Andrade, P. & Morais, P., 2001. Effects of alterations in fresh water supply on the abundance and distribution of Engraulis encrasicolus in the Guadiana estuary and adjacent coastal areas of south Portugal. *Ecohydrology and Hydrobiology* (3), 341–345.

Drobniewska, A., 2008. *Optimization of the use of sewage sludge for bioenergy production based on a large scale experimental willow plantation and mathematical model.* Ph.D. thesis, University of Lodz

Eagelson, P.S., 1982. Ecological optimality in water limited natural soil-vegetation systems. 1. Theory and hypothesis. *Water Resource Research.* 18, 325–40.

Gyawali, D., Allan, J.A., Antunes, P., Dudeen, A., Laureano, P., Luiselli Fernández, C., Scheel Monteiro, P.M., Nguyen, H.K., Novácek, P. & Pahl-Wostl, C., 2006. *PEU-INCO water research from FP4 to FP6 (1994-2006): a critical review.* [e-book]. European Commission. Available at http://ec.europa.eu/research/water-initiative/pdf/incowater_fp4fp6_rapport_technique_en.pdf [accessed 09 February 2011].

Hobbs, R.J., Arico, S., Aronson, J., Baron, J.S., Bridgewater, P., Cramer, V.A., Epstein, P.R., Ewel, J.J., Klink, C.A., Lugo, A.E., Norton, D., Ojima, D., Richardson D.M., Sanderson, E.W., Valladares, F., Vilà, M., Zamora, R. & Zobel, M., 2006. Novel ecosystems: Theoretical and management aspects of the new ecological world order. *Global Ecol. Biogeog.* 15, 1–7.

Junk, W.J., Bailey, P.B. & Sparks, R.E., 1989. The flood-pulse concept in river-flood-plain systems. *In: Dodge, D.P. (ed.) Proceedings of the International Large Rivers Symposium. Canadian Special Publication of Fisheries and Aquatic Science* 106, 110–127.

Kędziora, A., Olejnik, J., 2002. Water balance in agricultural landscape and options for its management by change in plant cover structure of landscape. *In: Ryszkowski L. (Ed.) Landscape ecology in agroecosystems management.* CRC Press. Boca Raton.

Kędziora, A., Olejnik, J. & Kapuściński, J., 1989. Impact of landscape structure on heat and water balance. *Ecology International Bulletin* 17, 1–17.

Kiedrzyńska, E. Wagner-Łotkowska, I. & Zalewski, M., 2008. Quantification of phosphorus retention efficiency by floodplain vegetation and a management strategy for a eutrophic reservoir restoration. *Ecological Engineering* 33: 15–25.

Krauze, K. & Wlodarczyk, R., 2010. *Drivers and pressures of city water resources – a risk-based decision support system for implementation of the Ecohydrology approach.* 5th SWITCH Scientific Conference 2010 "Sustainable Water management Improves Tomorrow's Cities' Health: achievements and way forward". 13–16 of October 2010, Lodz, Poland.

Mankiewicz-Boczek, J. Kaczkowski, Z. & Godowska, M., 2010. RNA/DNA ratio as an indicator of the impact of long-term accumulative contamination for the assessment of river degradation – a pilot study. *Ecohydrology & Hydrobiology* (2–4): xxx.

Mitsch, W.J.; Zhang, L.; Anderson, C.J.; Altor, A.E. & Hernández, M.E., 2005. Creating riverine wetlands: Ecological succession, nutrient retention, and pulsing effects. Ecological Engineering 25: 510–527.

Naiman, R.J. Decamps, H. Pastor, J. & Johnston, C.A., 1988. The potential importance of boundaries to fluvial ecosystems. *J. North A. Benthol. Soc.* 7: 289–306.

Ploskey, G. R., 1985. Impact of terrestrial vegetation and preimpoundment clearing on reservoir ecology and fisheries in the United States and Canada. *Food and Agriculture Organization of the United Nations, Fisheries Technical Paper* 258: 1–35.

Rodriguez-Iturbe, I., 2000. Ecohydrology: a hydrological perspective of climate–soil–vegetation dynamics. *Wat. Resour. Res.* 36, 3–9.

Ryszkowski, L. 2002. *Landscape Ecology in Agroecosystems Management.* CRC Press LLC. USA.

Tarczyńska M. & Zalewski M., 1995. Toksyczność zakwitów sinicowych jako wynik eutrofizacji zbiorników wodnych. *Gosp. Wodna* 4: 83–87.

Tundisi, JG. & Straskraba, M. eds., 1999. *Theoretical Reservoir Ecology and its Applications.* International Institute of Ecology. Brazilian Academy of Sciences and Backhuys Publishers, Leiden, 53–173.

UNEP., 2003. *Phytotechnologies. A Technical Approach in Environmental Management.* UNEP, Division of Technology, Industry and Economics. Freshwater Management Series No.7.

Urbaniak, M. Zieliński, M. Wesołowski, W. & Zalewski, M., 2008. PCBs and Heavy Metals Contamination in Bottom Sediments from Three Reservoirs of Different Catchment Characteristics. *Polish J. of Environ. Stud.* Vol. 17, No. 6, 941–949;

Urbaniak, M. Zieliński, M. Wesołowski, W. Zalewski, M., 2009. Sources and distribution of polychlorinated dibenzo-para-dioxins and dibenzofurans in sediments of urban cascade reservoirs, Central Poland. *Environ. Protec. Engineer*, No. 3, (w druku).

Wagner, I., da Silva Wells, C., Butterworth, J., & Dziegielewska-Geitz, M. 2010. *Reflection on the achievements and lessons from the SWITCH urban water management initiative in Łódź, Poland.* SWITCH city assessment paper [online] Available at: http://www.irc.nl/page/61360 [Accessed 08 February 2011].

Wagner, I., da Silva Wells, C., Butterworth, J., & Dziegielewska-Geitz, M. 2010. Łódź–city of water. In *SWITCH in the City: putting urban water management to the test.* Eds. Butterworth, J.A. McIntyre, P. & Da Silva Wells, C. The Hague, IRC International Water and Sanitation Centre.

Wagner, I. Izydorczyk, K. Drobniewska, A. Fratczak, W. Zalewski, M., 2007. *Inclusion of ecohydrology concept as integral component of systemic in urban water resources management. The city of Lodz case study, Poland.* SWITCH Scientific Meeting Proceedings. Birmingham, UK, January 2007.

Wagner, I., Izydorczyk, K., Kiedrzyńska, E., Mankiewicz-Boczek, J., Jurczak, T., Bednarek, A., Wojtal-Frankiewicz, A., Frankiewicz, P., Ratajski, S., Kaczkowski, Z. & Zalewski, M., 2009. Ecohydrological system solution to enhance ecosystem services: the Pilica River Demonstration Project. *Ecohydrology & Hydrobiology*, Vol 9, No 1, 13–39.

Wagner, I. Marshalek, J. & Breil, P. eds., 2008. *Aquatic Habitats in Sustainable Urban Water Management: Science, Policy and Practice.* Taylor and Francis/Balkema: Leiden.

Wagner, I. & Zalewski, M. 2009. Ecohydrology as a Basis for the Sustainable City Strategic Planning - Focus on Lodz, Poland. *Rev Environ Sci Biotechnol.* pp. 209–21; 7DOI 10.1007/s11157-009-9169-8.

Wolanski, E., Boorman, L. ., Chicharo, L., Langlois-Saliou, E., Lara, R., Plater, A.J., Uncles, R.J. & Zalewski, M., 2004. Ecohydrology as a new tool for sustainable management of estuaries and coastal waters. *Wetlands Ecology and. Management* 12, 235–276.

Zalewski, M., 2000. Ecohydrology. The scientific background to use ecosystem properties as management tools toward sustainability of water resources. Guest Editorial, *Ecological Engineering* (16) 1–8.

Zalewski, M., 2002. Ecohydrology–the use of ecological and hydrological processes for sustainable management of water resources. *Hydrol. Sci. J.* 47(5), 825–834.

Zalewski, M., 2006. Flood pulses and river ecosystem robustness [In:] Tchiguirinskaia, I., et al. 2006. eds. Frontiers in Flood Research, Kovacs Colloquium, UNESCO, Paris, IAHS Publication 305.

Zalewski, M. Brewinska-Zaras, B. Frankiewicz, P. & Kalinowski, S., 1990. The potential for biomanipulation using fry communities in a lowland reservoir: Concordance between water quality and optimal recruitment. *Hydrobiologia.* 200/201: 549–556.

Zalewski, M. Frankiewicz, P. & Berwinska-Zaras, B.. 1985. Factors limiting the growth and survival of brown trout introduced to different types of streams. *J. Fish. Biol.* 27: 59–73.

Zalewski, M. Janauer, G.A. & Jolankai, G., eds., 1997. *Ecohydrology: A New Paradigm for the Sustainable Use of Aquatic Resources.* UNESCO. Paris. Technical Documents in Hydrology 7.

Zalewski, M. & Naiman, R.J., 1985. The regulation of riverine fish communities by a continuum of abiotic-biotic factors. In: J.S. Alabaster, ed. *Habitat Modifications and Freshwater Fisheries.*, FAO UN. Butterworths, London. 3–9.

Zalewski, M. Tarczyńska, M. and Wagner-Łotkowska, I., 2000. Ecohydrological approach for elimination of toxic algal blooms in lowland reservoir. *Verh. Internat. Verein. Limnol.* 27 1–8.

Zalewski, M. & Wagner, I., 2005. Ecohydrology - the use of water and ecosystem processes for healthy urban environments. In: Special issue: *Aquatic Habitats in Integrated Urban Water Management.* Ecohydrology & Hydrobiology (4), 263–268.

Chapter 17

The Rotterdam approach: connecting water with opportunities

J.C.J. Jacobs

City of Rotterdam, Rotterdam, The Netherlands

ABSTRACT

The potential effects of climate change on large coastal cities have been addressed in scientific literature over the past 10–15 years. Climate change and sea level rise may exacerbate flood risks as well as river peak discharges. As a consequence, the vulnerability of infrastructure, people, nature and other economic sectors is increasing. Future climatic developments need to be taken into account in current urban and port development plans. It is important to reserve the space that will be required in the future for water storage, and to prepare projects that will be necessary to enhance protection against flooding and guarantee accessibility. In addition, design and construction concepts have to be adjusted both at the level of urban planning and at the individual building level. Taking these steps now will generate innovations that can also be marketed elsewhere in a later phase. This chapter discusses the water sensitive design initiatives that Rotterdam is taking, using climate change as a catalyst for urban design that is 'climate proof' but also using water to improve the overall urban environment.

The city of Rotterdam has decided to make the region 'climate proof', in other words, to create a system that is sufficiently robust and that can adapt in good time to changes as they occur.

Innovative applications in the area of water management will make the city more attractive. Examples include the construction of water plazas and floating structures.

Keywords: Climate change; adaptation strategy; delta cities

17.1 CLIMATE CHANGE: PREVENTION AND ADAPTATION

The potential effects of climate change on large coastal cities have been addressed in scientific literature over the past 10–15 years. Climate change and sea level rise may increase flood risks from storm surges as well as peak discharges from rivers. Furthermore precipitation patterns will become more extreme, leading to both extra flood risks and extended periods of aridity and high temperatures with a significant impact on society as a whole. As a consequence, the vulnerability of infrastructure, people, nature and other economic sectors is increasing.

In order to limit climate change, national and international climate agreements have been reached with a specific focus on the reduction of greenhouse gas emissions. For this reason, the Rotterdam Climate Initiative (RCI) was created. The objective of the RCI is to reduce CO_2 emissions in Rotterdam by 50% in 2025, as compared with 1990. The aspiration for the future is to become the first climate neutral port city in the world. To this end, Rotterdam participates in the Clinton C40 Initiative and plays an active role worldwide in making port cities climate neutral.

All these efforts notwithstanding, the city and the port will at the same time have to arm themselves against the present and future consequences of climate change. In general, socio-economic trends may further amplify possible consequences from future floods as more people are moving towards urban coastal areas and capital is continuously invested in ports, industrial centers and financial businesses. By 2040, more than 50% of the global population is expected to live in urban coastal areas. Adaptation to climate change and flood risks entails a broad portfolio of options. Adaptation strategies can be targeted to managing the probability of a flood or to minimizing its consequence. Furthermore, adaptation policies may address evacuation schemes, improved communication to civilians and companies and post event measures such as insurance and emergency relief.

To be better prepared for what the future might hold, Rotterdam has decided to pro-actively deal with the impacts of climate change. These challenges have been identified for the first time in the Rotterdam Water Plan (2007), which has led to the Rotterdam Climate Proof (RCP) programme. Although the future climate scenarios are under discussion, the city of Rotterdam has decided to make the region 'climate proof', in other words, to create a system that is sufficiently robust and that can adapt in good time to changes as they occur. That way, the city will be able to resist extreme conditions and continue to function as it functions today – or even better. Via the RCP programme on adaptation Rotterdam will maintain and strengthen the region's competitive advantages on the international stage while at the same time reducing its vulnerability to climate change and making use of the (urban planning) opportunities that come with it.

By recognizing that water could and should contribute to solve urban problems by upgrading neighbourhood quality, urban water management in Rotterdam was able to develop a climate adaptation strategy by creating urban development strategies that are sensitive to water issues. This approach of linking water infrastructure to urban development is known as 'Water Sensitive Urban Design' (WSUD) and reflects a new paradigm in which water and infrastructure design are integrally considered in the early stages of urban planning. The shift in urban water management in Rotterdam is part of a larger transition in Dutch water management from a sectoral and technological approach towards an integrated and interactive approach.

17.2 ROTTERDAM CLIMATE PROOF

Rotterdam is one of the safest port cities in the world. Maintaining this position requires considerable effort. At the same time, these efforts should be converted into opportunities. Flood protection in Rotterdam, for example, is realized in close connection with strengthening of the Rotterdam economy.

Climate adaptation and spatial development are inextricably intertwined in Rotterdam. This approach allows urban planners to create designs that effectively address the issue of climate change. At the same time, the necessity to invest in climate adaptation will create and sustain the momentum needed to enhance the city's attractiveness.

Adaptation to climate change is more than just a necessity for Rotterdam. Perhaps even more importantly, it offers an economic opportunity. This approach is characteristic of Rotterdam Climate Proof. Gaining knowledge and experience fast in combining water management and spatial planning will allow the corporate sector to build a powerful competitive advantage.

Thus the ambition will be realized on the basis of three guiding principles:

- Rotterdam will develop into and present itself on a national and international level as a leading centre for water knowledge and climate change expertise.
- Investments in climate solutions will enhance the attractiveness of the city and port for residents, companies, and knowledge institutes.
- Innovations and knowledge are developed, implemented, and marketed as an export product.

Innovative applications in the area of water management will make the city more attractive. Examples include the construction of water plazas and floating structures (Figure 17.1). These are appealing projects that will serve to set an example for other delta cities, particularly in view of the fact that Rotterdam is the lowest-lying delta metropolis in Europe and that it has one of the largest ports in the world.

The emphasis is on projects that will strengthen Rotterdam's position as a water city. Collaboration is key in this respect. Knowledge institutes and companies find a testing ground in Rotterdam for the challenging international climate change issues of our time. As the global demand for knowledge and solutions in the area of climate change adaptation is growing, Rotterdam has imperative reasons to highlight its qualities as a water knowledge city on an international level.

Figure 17.1 Floating pavillion

17.3 THE ROTTERDAM APPROACH

17.3.1 Three pillars

Rotterdam Climate Proof nourishes high ambitions. Their realization requires decisiveness, knowledge development, space to test and experiment, and new partnerships and alliances. For this purpose, Rotterdam has decided to adopt a clever way of combining policy and implementation, considerations and execution. Innovative solutions will enhance the safety and the quality of life in the city, while at the same time offering substantial economic potential for the entire region. Rotterdam thus becomes an inspiring example to other delta cities.

Knowledge: Rotterdam is developing into a leading water knowledge and climate city.

Trendsetting research in the area of theoretical and applied delta technology enables Rotterdam to create the basis for groundbreaking innovations and international collaboration. Within the national research programme entitled 'Knowledge for Climate', Rotterdam has been named a hotspot in this field. In concrete terms, this means that a fund of over 5 million euros is available for the 50% co-financing of research efforts. Accordingly, most of the knowledge projects are included in this programme.

Knowledge development also takes place through knowledge exchange. For this reason, an international knowledge network has been set up, called 'Connecting Delta Cities'. In addition, Rotterdam cooperates intensively with a large number of universities. New educational programmes focused on water management and delta issues are currently being developed, and students are given a chance to work for Rotterdam Climate Proof.

Actions: Rotterdam is developing into a global testing ground in the field of delta technology and adaptive building.

Investing in a climate proof city contributes to the realization of a safe, healthy, and attractive living environment and a strong economy for the residential and business communities. Innovations and knowledge are developed and applied for use in Rotterdam. Especially the *Stadshavens* district will be the recipient of substantial investments, for instance in terms of adaptive building and floating constructions (such as the floating pavilion), and the National Water Centre. This will help *Stadshavens* to become the showcase of Rotterdam for 'urban delta technology'.

Positioning: Rotterdam is an inspiring example to delta cities around the world.

Rotterdam positions itself as a delta city with a world port that approaches future climate change in an unassuming, safe and innovative manner. Not only on a local level but also on a national and international level, Rotterdam projects an image of innovation, reliability, and decisiveness. Rotterdam will act as a showcase, also by organizing conferences and taking part in major international events e.g. the 2010 World Expo in Shanghai.

17.3.2 Rotterdam adaptation strategy (RAS)

Climate change resilience in Rotterdam will first and foremost involve sustainable protection against flooding in the areas inside and outside the levees. In addition to flood management, the city will have to focus on other ways to prepare for the consequences of climate change as well, such as higher incidence rates of heat waves, increased heavy precipitation, groundwater salinization, changing options for transport by water, and increased volatility of groundwater levels. What is essential for an adaptive strategy is that it is implemented proactively and that it can be adjusted to changing circumstances. The Rotterdam Adaptation Strategy (RAS) clearly defines the measures that can be taken to climate proof the area.

The implementation of this proactive strategy requires knowledge and tools. In concrete terms, RAS consists of four clusters:

- Trajectory: Which specific objectives and milestones can be defined on the road to climate change resilience? This involves both a substantive definition and planning. Subprojects: developing a climate atlas and route planner.
- Specific activities: Keeping in mind the trajectory and the milestones to be attained, the question can be answered as to what project managers can do specifically to climate proof their areas. Subproject: developing a climate adaptation tool box.

- Procedure: This involves embedding climate adaptation into current work processes. Eventually, the issue of climate proofing will have to be included in the early phases of all spatial planning, policy-making, implementation and management. Subproject: incorporating climate adaptation.
- State of affairs: What is the current state of affairs for the projects of Rotterdam Climate Proof; which interim objectives have been achieved so far? How can they be made manifest? Subprojects: monitoring climate adaptation, and developing a climate barometer.

17.3.3 Five subject-related main themes

Within the RCP programme, implementation and knowledge-related activities are clustered into five subject-related main themes: flood management, accessibility, adaptive building, the urban water system, and the urban climate. Each theme was assigned a specific objective and a coherent set of products with a focus on knowledge development and implementation.

17.3.3.1 Flood management

Its location by the river Maas and close to the sea makes Rotterdam an outstanding example of a delta city. The city and the region are protected by a network of water defences. The levees, the dunes, and the *Maeslantkering* are crucial to flood management in Rotterdam and to the safety of the city. Climate change now poses new challenges for the city and the region. Dutch experts project a sea level rise between 65 and 130 cm by the year 2100 for the Dutch coast (compared with the level in 1990).

The Rotterdam region is a delta area where vast numbers of people live and work both inside and outside the levees. Not surprisingly, therefore, this region (officially called the Rijnmond-Drechtsteden Region) receives a lot of attention in the National Water Plan and the government's Delta Programme. At present, Rotterdam is a safe region, but in order to guarantee its safety in a changing climate, proactive measures are a necessity. After all, the prosperity of the economy and the attractiveness of the city depend on safety. It is, therefore, essential to invest in knowledge on water systems, the changes and fluctuations that may occur in these systems, and possible medium-term and long-term protection measures. Investing now will prevent excessive expenses in the long run.

In 2008 a special committee chaired by former Minister Veerman published its advice on the future of the Netherlands regarding water safety and water supply in times of climate change. These recommendations of the Veerman Committee will lead to a new Delta Programme for the Netherlands. Rotterdam plays a leading role in the elaboration of the regional Delta Programme Rijnmond-Drechtsteden. In collaboration with the government and a number of regional partners, various scenarios are explored for the future structure of water management.

Rotterdam wholeheartedly embraces the Veerman Committee conclusion that by 2050, the layout of the Rijnmond-Drechtsteden region should be designed to withstand the impact of storms and extreme river discharges, and considers this a condition for the further development of the city and port. Experts of the City of Rotterdam have been (and still are) involved in the elaboration of the initial ideas. And also the city council has pronounced political support. The mayor of Rotterdam is chairman of the regional steering committee. The main reason is that the ultimate solution for the area is of vital importance of the development of the city. The proposed alternative of a 'Lockable Open Rijnmond estuary' seems to offer great opportunities for spatial planning along the river and nature development.

Within the theme of 'flood management', research currently plays a pivotal role. The studies provide insight into the impact of climate change and into the effects of the development of adaptation strategies that will keep the region safe and protected against flooding. Following comprehensive consideration (of aspects including the social, technical, economic, spatial, and planning effects), these studies will serve as guidelines in the selection of the most appropriate, optimal adaptation strategy to ensure sustainable protection of the city and port.

One particular aim is to devise measures that will benefit several themes, and that allow for a combination of functions, for instance. Examples include innovative primary water defences in an urban context; levees that offer both protection and room for infrastructural, residential or recreational purposes. Yet another example concerns the installation of flexible flood defences with a lock-off valve to secure the Rijnmond region in times of heavy storms, while at the same time boosting the economic development of the areas outside the levees.

17.3.3.2 Accessibility

Proper accessibility is of vital importance for the economic development of Rotterdam. Proper functioning of freight transport and passenger transport by road, water and rail is crucial to the further development of the port, industry and services sector. But to what extent is the current transport infrastructure of the city and port climate change resilient?

Current climate change projections typically involve high temperatures, extended periods of drought or precipitation, extreme rainfall, storms, and a rising sea level. All of these characteristics may conceivably impact the network and the infrastructure of a city, in the form of subsidence, for instance, or melting asphalt, low water levels, or congested tunnels and bridges during extreme rainfall and storms. Bottlenecks already experienced during heavy downpours are flooded tunnels. Especially when the scarce underground passages under railways in Rotterdam are inaccessible, during rush hours the entire urban road system gets jammed. Moreover, one could wonder what the impact of stronger fluctuations in water levels will be on the navigability of the rivers. What needs to be done to preserve or expand the transport share of inland navigation? Which options are further available to strengthen passenger transport by water in the region? New mobility issues as a result of climate change will also need to be taken into account in this respect. The concept of living on water, for example, will obviously also increase the demand for transport by water. It is obvious that climate change will have a substantial impact on the layout of the infrastructure in Rotterdam.

17.3.3.3 Adaptive building

The expected sea level rise and the increased river discharge will increase the risk of future flooding of the areas outside the levees. As in all coastal cities, for Rotterdam, the development of waterfronts and the areas outside the levees is of significant importance. As this involves large investments, these areas will also be used more intensively. Accordingly, the social consequences and the potential damage due to emergencies such as flooding or high tide will increase substantially as well.

The current security strategy for the areas outside levees applied by the government, the water boards, and the municipality is focused predominantly on limiting the consequences of high water levels by strengthening levees and quays, overall heightening, and installing surge barriers and emergency water diversions. As insecurity on the consequences of climate change increases, however, this strategy becomes vulnerable. Confronting this insecurity requires a strategy that aims also to reduce the consequences of high tides. The concept of designing the areas outside the levees in a way that is resistant to high water levels is referred to as adaptive building.

For an optimal approach of adaptive building in the areas outside the levees, the subject is addressed from various angles, including the aspects of urban planning and construction method, access structure, mains services design, the organization of the open space, and legal measures (such as area-specific regulations on the method of construction and use). Key elements in this respect are permanent protection against flooding, and the development of a sustainable and attractive environment. Adaptive building thus offers opportunities for the realization in Rotterdam of new residential and business environments.

Rotterdam for instance, has ambitious plans to build floating urban districts. The *Stadshavens* district offers space for 1600 hectares of sustainable area development. By 2040, some 13,000 climate change resilient homes will be built here, approximately 1,200 of which will be built on water. In these floating districts, people will live on the water, get their groceries, work and spend their leisure time, for a large part on the water. This offers a wide ranch of research topics, e.g. how to optimize infrastructure, how to implement and secure public utilities etc. In the recently opened floating pavilion, a flexible drinking water connection is made but waste water is treated in the pavilion itself. In larger floating communities the idea of floating self-sufficient utility centres that can easily be plugged in is one of the many ideas.

17.3.3.4 Urban water system

Climate change already makes its impact felt in Rotterdam. Extreme downpours causing flooded streets and basements are becoming more common, and in the longer term, extended periods of drought can be expected as well. The capacity of the existing drainage systems and pumping stations is obviously insufficient to 'keep our feet dry' during extreme rainfall. Innovative alternatives for water storage, solutions for water collection during heavy downpours (emergency storage), and options to delay the discharge of rainwater are needed. Furthermore, in dry periods Rotterdam should have sufficient water of sufficient quality.

Developing a proper adaptation strategy requires more in-depth insight into the solution avenues available to prevent flooding and water shortages. The task at hand is to devise measures that will solve future bottlenecks while at the same time contributing to the economic strength and attractiveness of the city. For this reason, water storage in Rotterdam is integrated in the urban environment wherever possible. Against this backdrop, Rotterdam Climate Proof and the water boards collectively encourage the installation of green roofs, for example. During heavy rainfall events, these green roofs are a highly valuable solution for temporary water storage. In addition, the City is studying possible locations for the construction of water plazas (Figure 17.2). These water plazas – a Rotterdam invention – fill up in a controlled manner during heavy rainfall, preventing surrounding streets from flooding. In dry periods, the plazas serve as children's playgrounds.

Figure 17.2 Multi-functional water plazas

Other storage applications involve multifunctional car parks. The new car park near the Museumpark, for instance, will be equipped with an underground water storage facility. These and other projects are deployed by Rotterdam to increase the storage capacity of the existing sewerage system, but also help to preserve optimal quality of life in the city.

Guaranteeing the water quality in times of drought requires appropriate measures. If, at these times, saline water or water of poor quality is channelled in, this will reduce the quality of the urban water. That is why large-capacity water connections to the regional water system are currently being installed, so as to safeguard the supply of additional fresh water. One example concerns the Blue Connection in the South of Rotterdam that will be constructed in 2013. This connection is a waterway with green borders crossing the southern part of Rotterdam. The project aims to improve the quality of the environment, improve the water supply and thus guarantee a better water quality and is an attractive connection for rowing boats and canoes between the centre and the green areas outside the city. In addition, flexible water-level control in the drainage canals and pools is deployed for additional seasonal storage.

17.3.3.5 Urban climate

The urban climate has an impact on the daily operations in the city (Figures 17.3 and 17.4). At the same time, the urban layout and design influence the urban climate. Themes that could serve as examples include heat stress during more frequent heat waves, the extra impact of particulate matter on public health in times of aridity, or the increased frequency of occurrence of insect infestations. The need for outflow areas, shaded places and cool areas in the open space increases as summers grow hotter. Public gardens and parks should be climate change resilient and can contribute to heat stress mitigation. These themes will become even more important as building density intensifies.

The theme of the *urban climate* carries a focus on improving the quality of life and the attractiveness of the environment by adjusting the layout of the external public space and the design of the urban areas to climate change. The way to achieve this goal is by expanding the knowledge on the urban climate in Rotterdam and applying this knowledge in practice. In addition, investing in raising public awareness of the fact that the climate – even in our current situation – influences the personal and professional lives of the people in the city is essential. For example greening the roofs in Rotterdam: attractive info roofs have been applied to show the possibilities of green roof design, combined with public information on the Rotterdam subsidies for green roofs. Another example is the 'climate tram' in which trams and busses have measuring instruments on board. The results are published on the Rotterdam Climate Initiative website so people can see e.g. differences in temperature and the urban heat island effect can be shown.

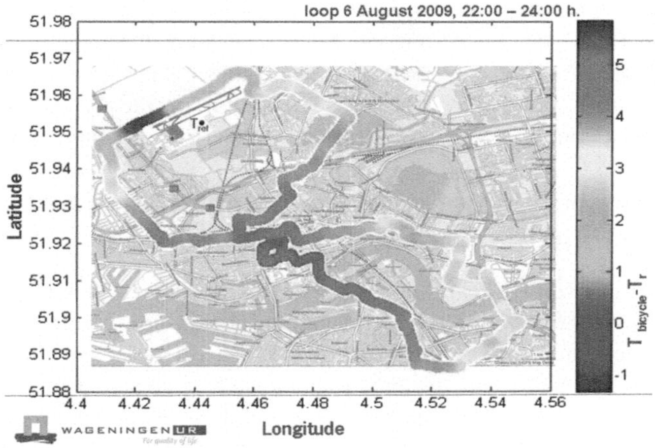

Figure 17.3 Rotterdam urban climate map

Figure 17.4 Bicycles with climate measuring instruments

17.3.4 Seven strategic projects

A number of strategic projects were defined within the programme as well. The focus in these projects is predominantly on image projection, showcasing Rotterdam as a model city and a living example, and the corresponding marketing. Of course, these projects will also contribute to knowledge development. These ambitions are embedded, for instance, in the Rotterdam pavilion at the Shanghai World Expo in 2010. Other strategic projects include the development of the National Water Centre, Connecting Delta Cities, the floating pavilion in Rotterdam, Smart Delta City, the implementation of the Water and Climate Marketing Plan, and the evolution towards the Rotterdam Adaptation Strategy 2.0.

17.3.5 Connection between the pillars, themes and strategic projects

The connection is self-evident: knowledge development and knowledge exchange are necessary for the innovative implementation of measures and the realization of a safe, robust city that demonstrates sufficient resilience (see Figure 17.5). 'Taking action' generates results and practical experience and creates a breeding ground for new, practical demand for expertise. This will allow Rotterdam to develop into a leading knowledge city and testing ground for climate-related issues and delta technology. Additional marketing activities and image projection will accelerate Rotterdam's national and international acknowledgement as a leading player in the water management and climate change arena. And, in turn, will increase the appeal of the city to knowledge institutes, corporations, residents and visitors.

As far as content is concerned, the pillars of knowledge and implementation are mainly aimed at the development of a set of instruments for Rotterdam's Adaptation Strategy (climate atlas, toolbox, barometer, route planners etc.) and subsequently very specifically at the five subject-related themes. The branding of Rotterdam as an advanced delta city and knowledge city for water and climate change management is embedded in the seven strategic projects.

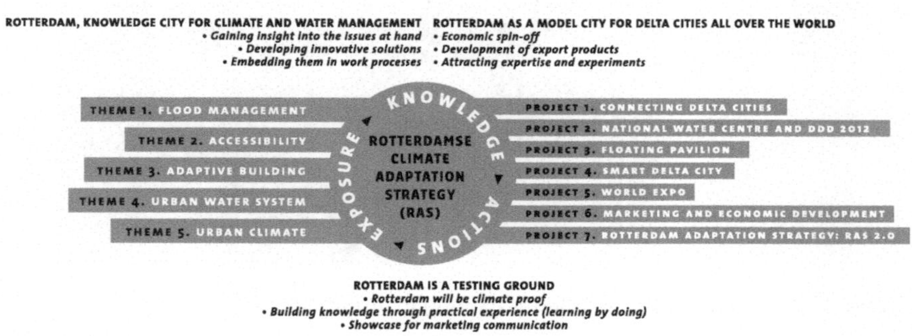

Figure 17.5 Rotterdam Climate Adaptation Strategy (RAS)

17.4 IMPROVING URBAN ENVIRONMENTS

As stated before, climate change measures and water solutions generate abundant opportunities to make the city even more attractive for residential, working, studying and entertainment purposes. Preventing the city from becoming unattractive due to risks of flooding, aridity or heat stress, can be combined with new environments, parks etc. We should always ensure that the measures are not only aimed at solving the problems, but also give an extra boost to the quality of life in the city. The creation of new canals for water storage is a good example. In an urban environment such as Rotterdam, these types of interventions and other traditional solutions such as levee strengthening are not sufficient. For the further development of the city it is, therefore, essential to include innovations such as green roofs, water plazas, underground water storage and cooling in urban design commissions. These innovations can really enhance the quality of the living environment.

Water disarms and binds people. You just have to look at the way children behave near a fountain. In general, people enjoy themselves at waterfronts, swim and play in river, ponds and canals. In adaptation projects in the city, citizens and different cultures come together. This can reinforce social ties and the sense of safety. The construction of green roofs, for example, can be set up as a neighbourhood project. Climate adaptation also generates new leisure opportunities such as floating parks. Water offers beautiful and safe possibilities for varied living environments, such as waterfronts and floating structures. In short, connecting climate adaptation measures with spatial opportunities reinforces the physical, economic and social pillars of the city.

REFERENCES

Aerts J. and Walraven A.M. (2008) Connecting Delta Cities, Sea Level Rise and Major Coastal Cities, Amsterdam, The Netherlands.

Francesch-Huidobro M. and Bakker M.H.N. (2009) A Multilevel-Governance Analysis of Climate Change Mitigation and Adaptation Measures of two C40 Delta Cities: Hong Kong and Rotterdam Compared. Hong Kong, The Netherlands.

Jacobs J.C.J. et al. (2007) Water Plan 2 Rotterdam. Rotterdam, The Netherlands.

Molenaar A. et al. (2008) Rotterdam Climate Proof, The Rotterdam Challenge on water and Climate Adaptation. Rotterdam, The Netherlands.

Molenaar A. et al. (2010) Rotterdam Climate Proof, Adaptation Programme 2010. Rotterdam, The Netherlands.

Veerman C.P. et al. (2008) Working Together With Water, A living land builds for its future. The Netherlands.

Chapter 18

Water challenges in building a sustainable city in the Middle East – a masdar perspective

Meghan Hartman[1], Ameena Kulaib[2] and Jay Witherspoon[1]
[1]*CH2M HILL, Abu Dhabi, United Arab Emirates*
[2]*MASDAR, Abu Dhabi, United Arab Emirates*

18.1 INTRODUCTION

Climate change has become the critical global topic in the 21st century. There is an immediate need to curtail and change current business practices and approaches towards self-sufficiency. Masdar City is Abu Dhabi's multi-faceted, multi-billion-dollar investment in establishing a city which will be a global example of innovative and sustainable technology implementation for energy, water, and waste management. The design and construction of the city have a zero-waste philosophy, making this initiative a global blueprint for urban sustainable development. Masdar's aim is to bring a solution to the marketplace that will change the global landscape. The implementation process is expected to benefit the development of all sustainable technologies.

In order for Masdar to reach its goals, technology delivery follows a triple-bottom-line approach which balances the social, environment, and economic feasibility of delivering a solution. Figure 18.1 illustrates how decisions are balanced based on their application, sustainability, and commercial needs. This approach provides robust criteria for decision making and technology selection. An acceptable technology must fall within the centre of each set of criteria in this diagram.

For Masdar to achieve its sustainability targets and goals there are four key strategies in place that enable this vision:

- Sustainable Infrastructure: From master-planning through detailed design Masdar will develop sustainable infrastructure. This will enable carbon neutrality to be achieved from operations including buildings, transport, water, materials, and waste.

- Sustainable Services: In order to make it easy and attractive for people to reduce their environmental impacts, Masdar will provide sustainable services and products such as sustainable textiles, food and consumer products.
- Sustainable Lifestyles and Behaviour: Awareness programs are required at Masdar to educate people about the sources of their impacts and creating a new culture of sustainability. This is necessary since not all lifestyle choices (e.g. choice of clothing purchased) can be influenced by the city design and facilities.
- Sustainable Facilities Management: This is critical from the city's environmental perspective. Teams will provide sustainable services on site on a scale not seen before: sustainability is a central focus in all estates management including lettings, sales, leasing, cleaning, maintenance, green infrastructure management, school management, health services management and so on.

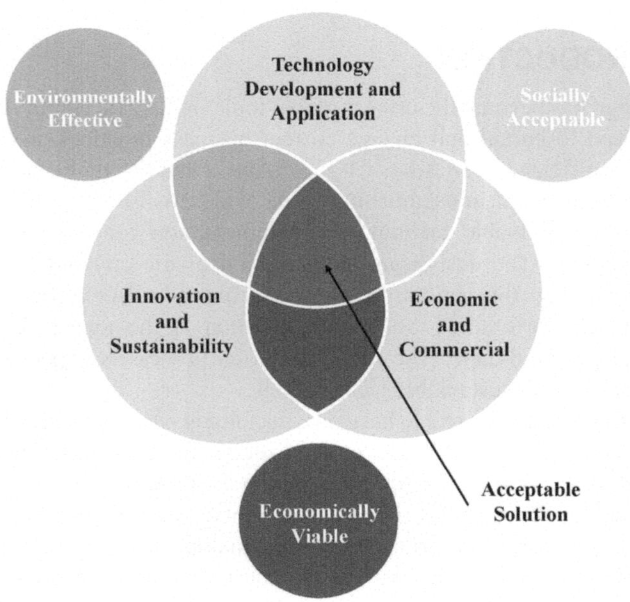

Figure 18.1 Triple-bottom-line considerations at Masdar

Masdar City covers an area of 7 square kilometers and is strategically located adjacent to the principal urban and regional transport infrastructure of Abu Dhabi.

Masdar's location is on the principal access road linking Abu Dhabi to its airport and to the neighboring emirate of Dubai. Masdar City is envisioned to be a technology hub that attracts clean technology businesses from around the world, who will become anchor tenants within the city. It is expected that the population of the city eventually will include 40,000 residents, along with an additional 50,000 commuters who work within the city. Construction of the Masdar Institute will be completed in fall 2010, while the first neighbourhood is scheduled for completion in 2014.

Through this initiative, Abu Dhabi aims to diversify its economy beyond oil and gas, and become a world leader in alternative energy. This includes a multilateral approach to energy; which will be provided to Masdar through photovoltaic (PV) installations, concentrated solar power (CSP), waste energy recovery, and both geothermal heat exchange and enhanced deep-rock geothermal energy. This multi-faceted approach is intended to supply both electrical and thermal energy for city usage. Similarly, a zero-waste philosophy will be adopted, with avoidance of landfill utilisation. The components of the waste stream will be separated for reuse in various technologies and or for possible energy production.

Sustainable transportation is also a major component of the initiative. Alternative transportation will be provided throughout the city by such means as light rail transit (LRT) or personal rapid transit (PRT). The LRT system is like any public transportation, with a main track through the centre of the city. The PRT system operates on an individual basis, but functions at a municipal level through the use of personal transports to transport people to specified destinations throughout the city, where each trip in unique.

To develop sustainable landscape design, Masdar invokes the traditional Islamic focus on open spaces, balancing the desire for shade and greenery with the limited water resources of the area, shown in Figure 18.2. The city is designed around several layered levels of varied use. The first layer is located 4.5 meters below grade, and houses the utilities for the entire city. The next layer at grade is designated for the PRT system. Residences and businesses, and open spaces are located above grade, with the first layer 5.5 meters above the PRT level. An example of infrastructure layering is shown in Figure 18.3.

Masdar City is in the centre of a major expansion zone of urban development, which may have significant impacts on the local potable water supply infrastructure. Accordingly, the city will adopt water minimisation, water recycling and re-use measures, and will also consider off-grid potable water sources. Masdar City adheres to the 'One Planet Living Principles of Sustainable Water' (Bioregional, 2008). The aim of this set of principles is to achieve a positive impact on local water resources by implementing water use efficiency measures, re-use and recycling; minimising water extraction and pollution. Fostering sustainable water and sewage management in the landscape and restoring natural water cycles.

Figure 18.2 Architectural drawing of Masdar City (Fosters and Partners, 2008)

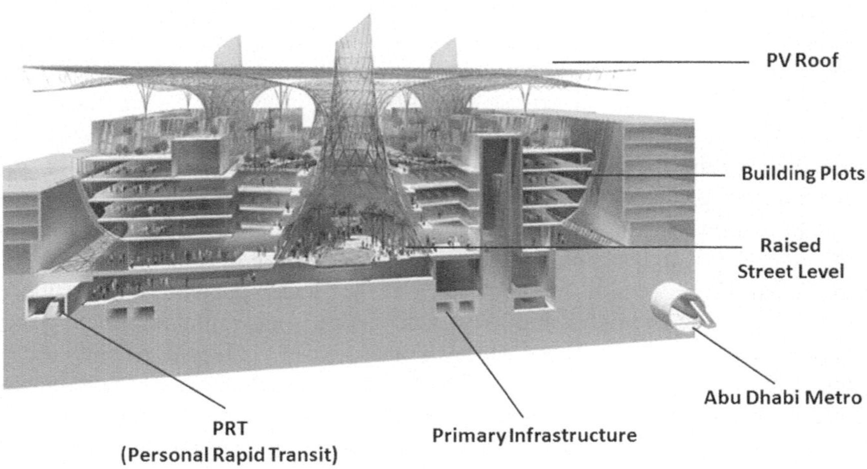

Figure 18.3 Cross-section of Masdar City and infrastructure layering

18.2 BACKGROUND

The supply and consumption of clean water is a critical concern for Masdar City which requires diligent management and planning. In the Emirate of Abu Dhabi, drinking water consumption is one of the highest in the world, averaging about 353 liters/person/day, compared with 120–150 liters/person/day in Western Europe. In the Emirate, the highest individual consumption rate is in villas, which average 525 L/person/day, mainly due to personal landscape irrigation. In published Abu Dhabi reports, residential consumption accounts for 22% of metered water use, from both groundwater and desalination (Environment Agency – Abu Dhabi, 2006). The high individual consumption rate is partly attributed to the Abu Dhabi government adopting a strategy of increasing production to meet demand rather than reducing consumption.

Water policy in the Emirate has principally emphasized increasing supply rather than efficiency gains through improved demand management. Water consumption is heavily subsidised and is projected to almost double by 2015 if water continues to be provided free of charge or at low cost to the consumers. Since the water tariff is low, this causes higher consumption due to the unrealized cost of usage.

Based on the growth of the new developments in Abu Dhabi, and without any change in behaviour, it is forecasted that by 2011 the water demand will outstrip the planned upgraded water supply which is due at just over 3640 Ml/d. By the time the Masdar development is fully constructed in 2016 it is estimated that the Abu Dhabi shortfall is estimated at over 910 Ml/d.

In the United Arab Emirates, there are no reliable perennial water resources available, and the Emirate is largely dependent on seawater desalination plants and groundwater extraction. Desalinated water accounts for 24 per cent of total demand and is costly: a cubic metre of desalinated water costs between Dh3 ($0.82USD) and Dh4 ($1.09USD) to produce, with consumers paying a fraction of the price. The groundwater consumption rate in Abu Dhabi is currently 26 times greater than that obtained from renewable sources. This over-extraction has led to a drop in the depth of the groundwater tables, which is not sustainable (Environment Agency – Abu Dhabi, 2007).

A strategic water plan for the Emirate of Abu Dhabi stressed that groundwater supplies are so over-tapped they would not be able to sustain current levels of use for more than 50 years. Sheikh Hamdan said the emirate's groundwater supply had fallen by 18 percent since 2003, while the consumption of water resources in the emirate exceeded their natural recharging capacity by 24 times (Todorova, 2009). Throughout the Emirate of Abu Dhabi, salinity is particularly high near the Arabian Gulf coast, and decreases towards the borders of Saudi Arabia and Oman, as shown in the predicted salinity concentration contour plot of the region in Figure 18.4 (GTZ, 2005). The area in which Masdar City is located is an area of brine quality groundwater.

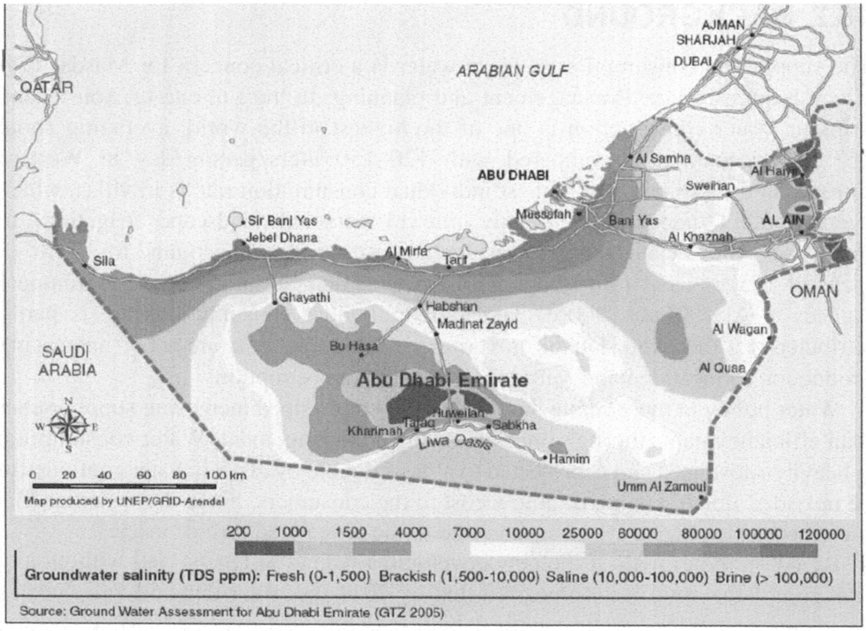

Figure 18.4 Salinity contours in the groundwater throughout the Emirate of Abu Dhabi (GTZ, 2005)

18.3 PROJECT DESCRIPTION

18.3.1 Sustainable water sources

Water in the Middle East is precious. For Masdar to achieve self-sufficiency, finding a potable water source is a formidable obstacle. A multifaceted approach to obtaining potable water sources has been planned for Masdar. A hydrogeological study was performed in 2008 that found on-site groundwater conditions with a TDS between 80,000 and 150,000 mg/L, which results in osmotic pressures too high for treatment with Reverse Osmosis (RO). Production of groundwater is considerably lower than the city's projected build-out water demand of approximately 10,000 m^3/day and shows an unsustainable draw-down in the aquifer.

Due to the elevated TDS, use the on-site groundwater is challenging in terms of treatment. Conventional desalination treatment methods are not applicable, and instead alternative methods have to be explored. Energy demand is a serious consideration within Masdar City, and treating hypersaline water requires large energy consumption. Technologies are currently being evaluated that pair desalinization with renewable thermal sources, such as concentrated solar power

(CSP), waste-to-energy, or enhanced geothermal energy. If both the technology, capital, and energy needs are met, a solution may be approached that utilises a sustainable portion of Masdar's groundwater.

These unique water and energy sources provide both a challenge and an opportunity for Masdar to capitalize on a diversified water portfolio. Success in such a challenging environment will provide a commercially viable approach to utilising and treating difference sources in less challenging environments. An investigative study is underway that looks at multiple sources: seawater, on-site ground water, sustainable withdrawal of off-site groundwater, and connection to the municipal water supply system. This approach combines technology innovation for varied qualities, while providing a robust and feasible solution with the lowest carbon footprint, cost, and environmental impact.

18.3.2 Water strategy

Masdar City will employ separate water conveyance systems to serve a variety of defined uses over distinct treatment levels. Planning for such intensive water management strategies requires the infrastructure development and technology to meet these ultimate goals. The preferred strategy is to undertake an indirect recycling scheme within the site which utilizes minimal amounts of desalinated water as a primary source. Greywater and blackwater streams are collected, treated to desired levels, and then returned to the city's water cycle for reuse and demand reduction. Figure 18.5 shows the planned water cycle for Masdar City.

Wastewater is a valuable resource to Masdar for residual utilization and recycled water. Greywater is defined as spent water from sinks, showers, and wash basins possessing low BOD and pathogens. This excludes spent water from toilets, dishwashers, and kitchen sinks. Blackwater from toilets, dishwashers, and kitchen sinks is often considered wasted water. At this point in time, greywater and blackwater will be collected together. Due to the lack of head from gravity conveyance, Masdar City plans to construct a vacuum sewer system to convey wastewater to a central treatment plant. Temporary membrane bioreactors (MBR) are being utilised for treatment, while technology innovations are underway for future construction phases and piloting. After treatment, wastewater will be used for landscaping, district cooling, and potentially indirect reuse for aquifer recharge.

All biosolids from blackwater treatment will be reused either by composting or energy production via combustion (waste to energy). There is also a goal to utilize brine from on-site dewatering and if potable water desalination is pursued. The desire would be to create markets for salt use and constituents present in brine. The financial viability and potential end uses are still yet to be determined for brine stream recovery.

Native plant species will be used throughout the city in order to further reduce any irrigation needs. These plants are adapted to the desert climate and require minimal amounts of water. Use of native species will also curb the high irrigation demand that is seen throughout Abu Dhabi. These plants can also be enhanced by using specialised methods such as drip and micro-spray irrigation.

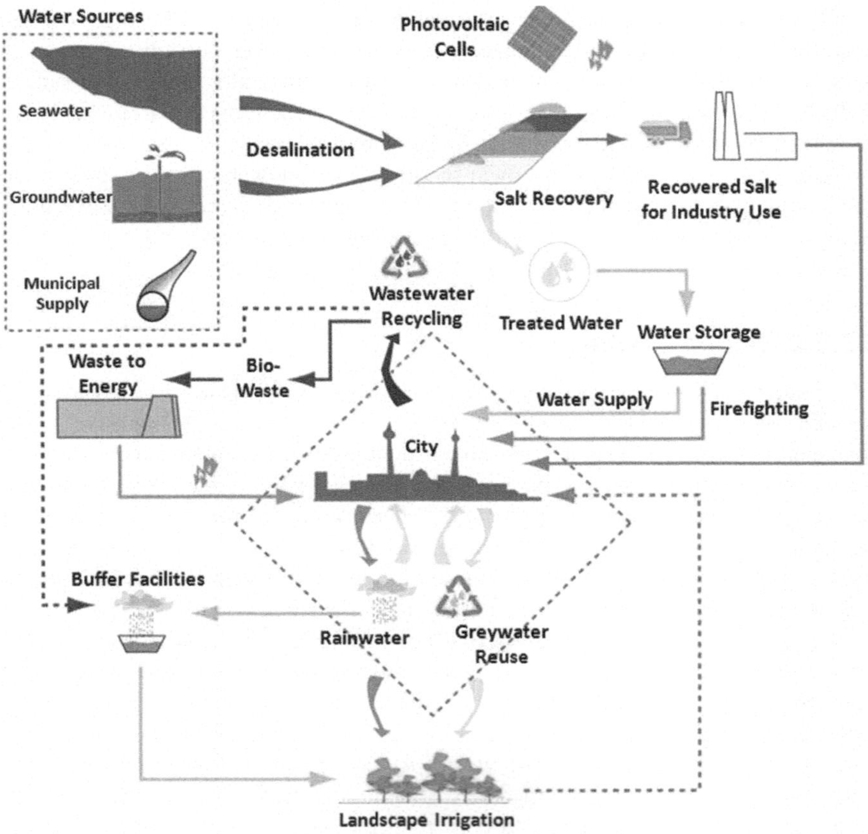

Figure 18.5 Masdar Water Cycle, (Fosters + Partners, 2008)

Masdar will also adopt water efficiency measures that minimise consumption through a combined approach of technology and cultural change. As mentioned, the Emirate of Abu Dhabi consumption is approximately 353 liters/person/day. In Masdar, it is envisioned to reduce consumption drastically through education measures, installation of water efficient fittings, and water recycling. The first phase would implement water efficient fittings, which provide the technology side to reducing consumption. Following technology implementation, an education program is introduced that initially leads the population towards transitional water-saving behaviour and eventually enlightened behaviour. The end result being a drastically reduced net potable water demand required for desalination. These consumption and demand numbers are listed in Table 18.1.

Table 18.1 Masdar water consumption with incremental reduction measures

Benchmark	Consumption (Liters/person/day)
No Water Savings (Abu Dhabi Average)	353
Best Practice Fittings, Conventional Behaviour	180
Best Practice Fittings, Transitional Behaviour	145
Best Practice Fittings, Enlightened Behaviour	105

18.4 REGULATION AND INFRASTRUCTURE DESIGN

While Masdar City envisions a completely sustainable infrastructure, the surrounding area still operates using best-practices of the region. Accordingly, the strategies implemented in Masdar are expected to be transferred to the rest of the region once proven success has been attained.

Masdar City's vision includes reuse of all water streams, but in planning stages, reuse regulations did not exist in the Emirate. Although many developers throughout the UAE recycle wastewater, the practice was unregulated. The Regulation and Supervision Bureau (RSB) has released consultation papers throughout 2008 and 2009 to try to understand public opinion on reuse of treated wastewater streams. In 2010 the RSB passed a wastewater reuse regulation for implementation in Abu Dhabi. Masdar has undertaken piloting programs to demonstrate achievable quality in treated greywater and blackwater. The piloting effort has been in conjunction with RSB as to provide informative and transparent testing results.

The overall design of Masdar City is unique in that it embraces traditional Arabic methodologies. This includes narrow streets for shading and allowing wind to naturally cool the city along courtyards and open markets for community usage. Innovation in design is also translated to utility development. The city has been designed so that flexibility, so maintenance and upgrades can be achieved easily.

With respect to water supply, flexibility is implemented through a city-wide utility trench or 'undercroft'. All water conveyance piping will be laid in the undercroft. This allows for ease of access when maintenance is required, leak detection, and efficiency in implementing system upgrades. Figure 18.3 shows the various considerations within the city along with utility infrastructure in lower layers.

Each building is also designed to have three unique water streams enter and exit a unit. Specifically, potable water enters the building for personal consumption and bathing. Treated wastewater also enters the building for toilet flushing use. Spent wastewater is passed out of the building for treatment and reuse. The water treatment and wastewater treatment plants will be centrally located due to the large central demand for each of these treatment streams.

18.5 TECHNOLOGY INNOVATION

Masdar envisions a technology roadmap to attract and develop innovative water technologies. This involves implementing the best innovative technologies of today, while developing and evaluating technologies that will eventually lead the market tomorrow. This vision requires a balance between delivery and technology where flexibility is built into infrastructure. For the first phases of Masdar City construction, water treatment will be achieved by utilising existing Abu Dhabi utilities while implementing temporary packaged treatment plants.

After the city transitions into the operation stage, testing of innovative technologies will remain. Specifically, a cost-benefit, financial and risk analysis will be performed for each of the technologies implemented. If a technology shows limited success for the cost required for its use, it will be re-evaluated. The city design builds upon flexibility and ability to enhance technology based on the feedback from the testing program.

On-site technology advancement will also be elevated through the unique partnership with the Masdar Institute. The Institute was developed in cooperation with the Massachusetts Institute of Technology (MIT). The Masdar Institute endeavours to become a centre of high-caliber renewable energy and sustainability research, capable of attracting leading scientists and researchers from around the world. The aim is to develop a pool of highly skilled scientists, engineers, managers, and technicians capable of accelerating the development of locally-developed technology and enterprise in the region and globally.

One of the largest obstacles facing technology is the delivery and implementation. In order to achieve this goal, Masdar has set out a process by which to identify, evaluate, study, test, and eventually implement the best practice to fit desired needs. Figure 18.6 shows a schematic view of the stages involved in the process for technology evaluation.

The technology evaluation process will involve the following testing stages:

1. Identifying technology gaps
2. Modeling of system integration and needs
3. Quantifying utility supply and demands
4. Evaluating a the long-list of technology options with the goal of generating a short-list
5. Performing a lifecycle analysis
6. Creating a business case for selected technology, and
7. Measuring the Key Performance Indicators (KPIs) for determining success

This roadmap will be followed to narrow a large list of technologies to identification of the optimal site-specific solution, preliminary studies, piloting studies, and full-scale implementation. This process will consider the triple-bottom-line methodology, but is directly applied to the business principle of delivery. This is in contrast to typical business decisions in which low cost overrides environment impacts. In order to achieve Masdar's ideals, a business case will be pursued only if it can fill a needed gap or advancement in technology or efficiency.

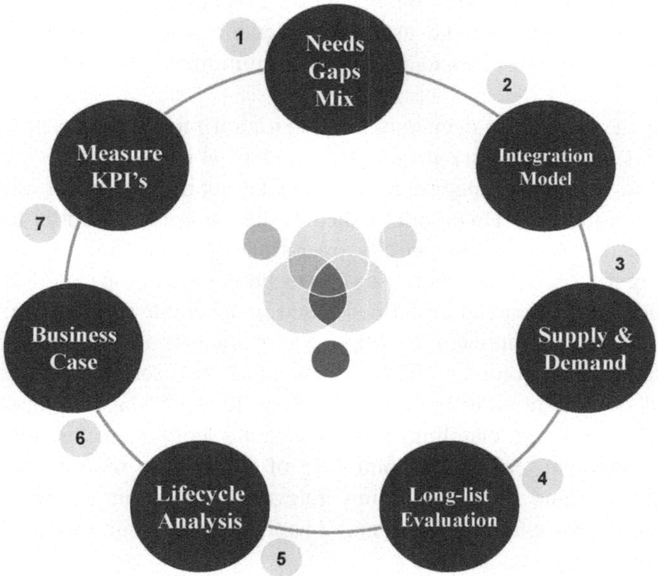

Figure 18.6 Technology Selection for Testing and Implementation

An example of roadmap process application was in evaluating potential greywater treatment technologies. Masdar wanted to determine the applicability of a decentralised or centralised greywater options along with measuring treated water quality. This led to an 'Open Call for Technologies' or an Expression of Interest which was issued in January 2009. From the responses received, an evaluation was performed on the long-list of technologies and their viability in meeting Masdar's goals. Included in the evaluation were triple-bottom-line considerations including social acceptability (aesthetics, etc.), impact on environment, and cost considerations. From this, a short list of potential technologies was developed that led to a business case. Monitoring and evaluation will be ongoing for the selected pilot projects that will be measured against initial gaps and KPIs.

18.6 SYSTEM INTEGRATION

Achieving Masdar objectives requires integration of sustainable water systems and technology into the overall functionality of the city. This includes energy generation, construction, and operation. Optimizing water supply and use conditions for city needs will certainly be largest significant challenge as city construction continues and operation begins. This challenge will be addressed through the use of systems modeling and other tools. With these tools, Masdar will be able to optimize sources of supply and level of treatment required for the water with the use of recoverable resources, energy demands, construction timing, waste management, and minimization of the city's carbon footprint.

To assist in technology decisions, an integration model was used to simulate scenarios and optimize utility needs. The model, which was previously used in earlier phases, shows the integration of the water supply, treatment and use with other feature of the city. This model indicates the variety of potential inter-related effects of the different sectors on city operations. A schematic of this model is shown in Figure 18.7. For water, varying technologies can be modeled and the optimal process selection can be refined based on available information.

Another critical component to Masdar's operations is the Information and Communication Technologies (ICT) system. The ICT services in Masdar City will make a significant contribution to the daily lives of Masdar citizens. It adds tangible value through enabling more efficient deployment of services (e.g. demand-management, enhanced monitoring of facilities to reduce consumption), and intangible value (e.g. providing relevant and accurate information to citizens, enhancing entertainment, and keeping citizens informed about their environmental progress agenda).

Masdar's sustainability initiatives require extensive ICT support – from the monitoring of consumption across the city, to integrating automatic demand management into the built environment through a wide range of sensors and intelligent appliances (e.g. temperature monitoring driving cooling consumption, light meters and motion sensor directing the use of powered lighting).

In keeping with the philosophies of sustainability and zero-waste, Masdar recognizes the integration of water management with other resource sectors, including energy and solid waste. This requires consideration and integration of all systems. Implementation decisions in the city cannot be solely made my one utility group. For all water treatment systems, robust investigation needs to evaluate the energy consumption and waste produced.

18.7 CONCLUSIONS

Masdar presents a unique opportunity in its ability to demonstrate innovative water technologies and concepts. The construction of the city focuses on minimisation principles while pairing with innovative technology. This vision is delivered by following the triple-bottom-line approach in both water and technology selection.

Masdar's water strategy requires innovation in balancing water sources that can be utilised sustainably while still achieving energy and water quality targets. The

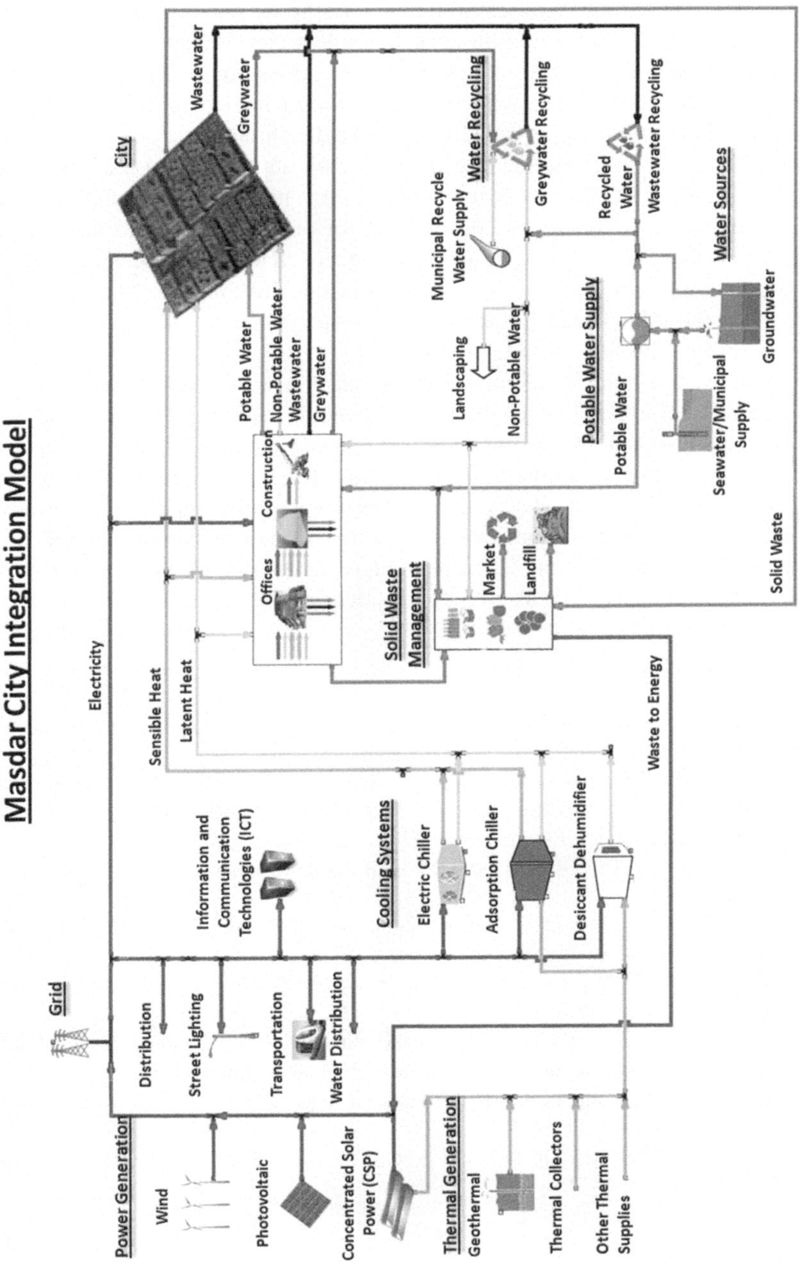

Figure 18.7 Integration model snapshot of citywide services

water cycle within the city focuses on maximizing reuse potential while coupling reduced potable water consumption. The multiple water streams within the city (potable, wastewater, reuse), provides multiple options for reuse potential. The city infrastructure is designed with flexibility for maintenance access, but also allows for technology to be upgraded with future developments.

Masdar City aims to make the development environmentally sustainable along with cost sustainable. Currently in the UAE, there is not a governmental structure in place that accounts for 'green construction' or provides credits to developers. Masdar has undergone several scenarios to ensure that rental prices are in line with Abu Dhabi prices, but does not inhibit sustainability targets. Masdar is also in talks with the Abu Dhabi government to develop frameworks to account for triple-bottom-line installations and credits.

Technology development requires a thorough process for selection. From gap identification, a multi-step evaluation is delivered that utilises modelling and technology analysis that may eventually lead to study or implementation.

From the groundbreaking work being developed at Masdar, the lessons learned and challenges will aid in delivering future sustainability goals and projects. Individual projects or indicators can be repeated in new locations or used as a template for future application. The knowledge gained from Masdar is critical in curbing climate change while also providing ideals for others to follow.

REFERENCES

Bioregional (2008). Masdar Outline Sustainability Action Plan, Benchmarked Against One Planet Living. *Bioregional Development Group.*

Deutsche Gesellschaft für Technische Zusammenarbeit (GTZ), Abu Dhabi National Oil Company (ADNOC), Environment Agency Abu Dhabi (EAD) (2007). Groundwater Assessment, Abu Dhabi. *Environment Agency Abu Dhabi (EAD).*

Environment Agency – Abu Dhabi (2007). State of the Environment. *Environment Agency Abu Dhabi (EAD).*

Environment Agency – Abu Dhabi (2006). Statistics 2006, Abu Dhabi Emirate Water Resources. *Environment Agency Abu Dhabi (EAD).*

Foster + Partners (2008). Masdar Detailed Masterplan. *Abu Dhabi Future Energy Company, Masdar Development Project.*

Regulation Supervision Bureau (2009). Wastewater Residual Reuse Regulations 2009 Public Consultation. *Regulation Supervision Bureau.* Publication CR/T06/004.

Regulation Supervision Bureau (2007). Annual Report 2007. *Regulation Supervision Bureau.* Publication ER/P02/016.

Todorova, Vesela (2009). Abu Dhabi Faces Water Crisis. *The National*, Abu Dhabi, United Arab Emirates, March 22.

IWA Publishing's authorised EU representative for General Product Safety Regulations is Diane D'Arras, 15 rue Duret, 75116 Paris, France, e-mail: safety@iwap.co.uk.

Printed and bound by CPI Group (UK) Ltd, Croydon, CR0 4YY
12/05/2026
02108883-0001